Ecocriticism of the Global South

Ecocritical Theory and Practice

Ecocritical Theory and Practice highlights innovative scholarship at the interface of literary/cultural studies and the environment, seeking to foster an ongoing dialogue between academics and environmental activists. Works that explore environmental issues through literatures, oral traditions, and cultural/media practices around the world are welcome. The series features books by established ecocritics that examine the intersection of theory and practice, including both monographs and edited volumes. Proposals are invited in the range of topics relevant to ecocriticism, including but not limited to works informed by cross-cultural and transnational approaches; postcolonialism; posthumanism; ecofeminism; ecospirituality, ecotheology, and religious studies; film/media and visual cultural studies; environmental aesthetics and arts; ecopoetics; ecophenomenology; ecopsychology; animal studies; and pedagogy.

Recent Titles

Ecocriticism of the Global South

Edited by Scott Slovic,
Swarnalatha Rangarajan,
and Vidya Sarveswaran

LEXINGTON BOOKS
Lanham • Boulder • New York • London

PN
98
.B6
.E38
2015

Published by Lexington Books
An imprint of The Rowman & Littlefield Publishing Group, Inc.
4501 Forbes Boulevard, Suite 200, Lanham, Maryland 20706
www.rowman.com

Unit A, Whitacre Mews, 26-34 Stannary Street, London SE11 4AB

British Library Cataloguing in Publication Information Available

Library of Congress Cataloging-in-Publication Data Available

ISBN 978-0-7391-8910-8 (cloth : alk. paper)
ISBN 978-0-7391-8911-5 (electronic)
ISBN 978-1-4985-1588-7 (paper : alk. paper)

♾™ The paper used in this publication meets the minimum requirements of American National Standard for Information Sciences Permanence of Paper for Printed Library Materials, ANSI/NISO Z39.48-1992.

Printed in the United States of America

Contents

Acknowledgements

This volume is the second half of a two-book "mini-series," the first of which, titled *Ecoambiguity, Community, and Development: Toward a Politicized Ecocriticism*, appeared at the beginning of 2014. We started working together on this project in December 2010 during a conference on ecocriticism at Tamkang University in Taiwan and wish to thank the organizers of that meeting, especially Peter Huang and Robin Tsai, for providing an exciting seedbed for our initial discussions. We also wish to express our gratitude to the Department of Humanities and Social Sciences at the Indian Institute of Technology Madras for facilitating our collaboration in Chennai in August 2012, during which we pulled together the specific plans for these two books, still thinking of them at the time as a single tome devoted to calling forth the ecocritical voices of the developing world. Our plans took on new clarity and concreteness in the summer of 2013 when we teamed up with Douglas Vakoch, editor of the new Ecocritical Theory and Practice Series for Lexington Books, and editor Lindsey Porambo from Lexington Books, to orchestrate the details of what quickly morphed into two separate books. We appreciate the interest and practical support Doug and Lindsey—and other colleagues at Lexington Books—have contributed to this project.

This work emerges from our belief that in every corner of the earth, wherever human beings reside or transit through, there are unique and important visions of the relationship between human culture and the natural world. In many cases, these perspectives result from colonial, postcolonial, and sometimes neocolonial conditions. Frequently, indigenous cultural practices and modes of expression inform the methodologies and purposes of ecocriticism in specific regions. Our hope in this Ecocriticism of the Global South Project is emphatically to welcome hitherto unheard voices—or voices muffled for various reasons—to the international conversation about who we are as a species and how best to understand our presence on this beautiful and troubled planet. We thank our contributors for their participation in this project.

Introduction

Ecocriticism of the Global South

This book takes as its starting point a concern that mainstream environmental discourse, particularly in the field of ecocriticism, has until recently been institutionally and epistemologically centered in the United States and the United Kingdom, while many of the actual ecological and social concerns that inspire such discourse are grounded in provincial realities. In the middle of the past decade, ecocritics such as Elizabeth DeLoughrey, Renée Gosson, and George Handley in their edited volume *Caribbean Literature and the Environment* (2005) and Rob Nixon in his essay "Environmentalism and Postcolonialism" (2005) began powerfully extending the social consciousness of environmental justice ecocriticism to postcolonial contexts throughout the world, recognizing, as Nixon put it, that environmentalism tended to be "irrelevant and elitist" (716) and insensitive to the fundamental intersection between "the ethics of place and the experience of displacement" (721). With the emergence of postcolonial ecocriticism, a field that has taken off in the years since its initial articulations in the above-mentioned works and now includes numerous collections and monographs, the impossibility of disarticulating social and ecological concerns has become plain.

More recently, Nixon has emphasized the relative imperceptibility of eco-social degradation in the developing world in his powerful work *Slow Violence and the Environmentalism of the Poor* (2011), adapting the "environmentalism of the poor" rubric from Ramachandra Guha and Juan Martinez-Alier to suggest that there is a distinct environmental epistemology and ethos associated with the experience of citizens in the Third World. For the most part, however, existing studies—both monographs and edited collections—operate from the point of view of First-World scholars, speaking *on behalf of*

1

subalternized human communities and degraded landscapes. What has only recently begun to happen in ecocriticism postcolonial or otherwise, though, is the concerted effort to let "the empire write back," as Bill Ashcroft, Gareth Griffiths, and Helen Tiffin phrased it in their classic work of postcolonial scholarship back in 1989. *Ecocriticism of the Global South* seeks to give voice to the empire, to grant ecocritical agency, if you will, to scholars living in, coming from, or in other ways deeply familiar with regions of the world (even in the Northern Hemisphere) that have traditionally been un- or under-represented in the halls and pages of ecocritical scholarship, one of the fastest growing and most socially urgent branches of the humanities. This volume extends the work we have begun in our 2014 collection *Ecoambiguity, Community, and Development: Toward a Politicized Ecocriticism*, which offers twelve chapters similarly seeking to write back to the world's centers of political, military, and economic power, expressing views of the intersections of nature and culture from the perspective of developing countries.

In 1980, German chancellor Willy Brandt proposed a distinction between the more developed regions of the world and the less developed regions, between "North" and "South." The so-called Brandt Line advocated economic development of Third World societies to the south of the thirtieth parallel, which separates North and Central America, passes above Africa and India, but does not include Australia and New Zealand which are considered de facto members of the North. Both "rich" and "poor" nations today fall on either side of the Brandt Line. *The Brandt Report* (1980), though best known for identifying economic disparities across the planet, also issued an apocalyptic warning of "a scorched planet of advancing deserts, impoverished landscapes and ailing environments" that our successors will inherit, if we do not pay attention to the increasingly complex relationships between our societies and the spaces that we inhabit. This book highlights the diversity of voices and perspectives that exist "South" of the Brandt Line rather than attempting to present a singular view of "development discourses" applied to the peoples and places of the Global South. By drawing forth the voices of literary scholars attentive to the particular cultural and environmental conditions of the Global South (including continents such as Africa and South and Central America, such regions as Northern Ireland, the Persian Gulf, and Eastern Europe, and such subcultures as Maori New Zealand, islanders from the Pacific Ocean, and Native Americans), we hope to fill a conspicuous gap in contemporary international ecocriticism.

In attempting to define the term "Global South," we acknowledge three loci of enunciation:

1. The Global South as an entity that has been invented in the struggle between imperial global domination and decolonial forces which resist global designs through their emancipatory articulations.

2. The Global South as a geopolitical concept replacing the "Third World" after the collapse of the Soviet Union. From this perspective, the Global South is the location of underdevelopment and emerging nations that need "support."
3. A third trajectory is the Global South within the North enacted by massive migration from Africa, Asia, South-Central America, the Caribbean, and the "former Eastern Europe." The "manywheres" of the "global" South acts as a springboard for critical imaginations of environmental consciousness and race.

The imbrications of the social and the ecological in the Global South pose important questions for ecocriticism. The complex relationship between literary texts from developing regions of the world and the threatened environments that produced them suggests the need for new practices of reading both the texts themselves and the place of humans in nature more generally. *Ecocriticism of the Global South* examines the close relationship between "ecology and the politics of survival" (to use the title of Vandana Shiva's 1991 book). It takes on the task of articulating the socio-ecological plight of the world's poor by drawing attention to the fact that uneven patterns of neoliberal development in the Global South threaten the millions who depend upon access to natural resources for their survival.

Building upon the twelve chapters published in *Ecoambiguity, Community, and Development*, we offer fifteeen new ecocritical chapters in this book. Many of these authors are distinguished postcolonial critics and/or ecocritics, and others are up-and-coming scholars from under-represented regions of the world in the field of ecocriticism (or living in North America or Western Europe and specializing in ecocritical approaches to the literature/culture of under-represented regions). We hope this collection illuminates the most trenchant aspects of the field of ecocriticism from a "southern perspective," even if a few of the "speakers" here are not themselves from the regions they are representing in this book. This book represents one of the most thoroughly multicultural examples of ecocriticism to date, both in the subjects treated and in the contributors whose voices are represented—in a sense, this multiculturalism in itself fundamentally represents the future of the field.

Amitav Ghosh's 2004 novel *The Hungry Tide* is one of the literary texts most frequently studied from the perspective of postcolonial ecocriticism. Standard approaches to this important text include Rob Nixon's examination of the disposability of displaced people who live in a rural and marginalized place such as the Sundarbans islands in between India and Bangladesh and Lawrence Buell's emphasis on the threatened ecology of this region. In a sense, this literary work and the sometimes divergent critical approaches scholars have applied to it represent the potential tensions between third-wave ecocritical attention to transcultural interactions in the environmental context and so-called first-wave ecocritical emphasis on literary views of ecological paradigms in the meta-human world. However, there is no intrinsic separation between cultural and ecological concerns. This is, as much as anything else, the first premise of postcolonial ecocriticism. In her study of Ghosh's novel, Priya Kumar confronts the potential tension between third-wave and first-wave ecocritical methodologies head on, ultimately showing that this text "makes clear that environmentalism does not have to be antihuman as the very spectrum of what counts as environmentalism has expanded with the increasing visibility and credibility of environmental justice movements." This chapter clarifies as well the essential consonance between environmental justice ecocriticism and postcolonial ecocriticism, both of which are modes of analysis geared to illuminate and resist social injustice in the context of environmental degradation and resource extraction.

We remain in South Asia with Sharae Deckard's study of Sri Lankan fiction, where the focus is on "war ecology" and, in particular, on the twin degradations, social and ecological, emerging from the Sri Lankan civil war. As Deckard puts it plainly, "no environments are ever ahistorical, untouched, or idyllic." Even now, nearly two decades following William Cronon's vigorous argument in "The Trouble with Wilderness" (1996) that "The flight from history that is very nearly the core of wilderness represents the false hope of an escape from responsibility, the illusion that we can somehow wipe clean the slate of our past and return to the tabula rasa that supposedly existed before we began to leave our marks on the world" (80), there persists an impulse, at least in Western culture, to decouple the fallen, fractious, disquieting course of human history from The Alternative: a place of retreat that we name "nature." But when nature is viewed through the lens of a culture fraught with violent conflict, it becomes what Deckard calls "a militarized zone," its organic systems transformed into "war ecology." Exceptionalist myths, such as pastoral retreats from urban and first-world stress and disorder, evaporate. This is what happens in ecocriticism of the Global South—or in the ecocritical reading of literature from the Global South. The innocent, distant retreat from the industrial decadence of the Global North is revealed as impossible fantasy.

Even when the Global South is not the site of outright war, there are frequently social processes at work that produce human and environmental suffering. Zhou Xiaojing illumunates such processes in her study of Chinese poet Zheng Xiaoqiong, who uses literature as a vehicle for describing the lives of migrant workers, drawn into exile from their home villages in order to live and work in industrial hives in eastern China. Even as China leaps forward to become, at present, the world's second largest economy, there continues to be a Third World underclass in that country, a subaltern society that exists to serve the monied class both domestically and internationally.

Insofar as it sheds light on the psychological and ecological traumas created by worker displacement and the processes of China's industrial juggernaut, ecocriticism serves as a vital tool of counter-friction against the machine, as Henry David Thoreau would have put it. Ecocriticism resists precisely the kind of human and environmental exploitation that Zheng's poetry describes.

Lest we feel a tendency to pull back and say, "Such problems exist over there, far away from here. They're none of our business," Christopher De Shield's study of literary isomorphism in literature of the Malayan and Caribbean archipelagos contracts distant regions of the world and demonstrates how colonial and touristic discourses and material practices are shared across the planet. Relying on the discourse of resistance offered by Martiniquan theorist Édouard Glissant, De Shield promotes the importance of at once recognizing the unique condition of individual places and the common patterns of oppression and struggle among colonized environments and societies. He refers to this tension between specificity and relationality as "almost a disavowed dialectic within postcolonial studies." But in order for a project like this book to work well, offering an eclectic treatment of locales and literatures from throughout the world, precisely this dialectic must be, to some extent acknowledged. We must agree to adapt discussions of particular locations to our thinking about other landscapes and societies, perhaps even our own.

From Malaysia and the Caribbean, we turn to New Zealand, where local scholar Charles Dawson provides an uplifting story of the conjunction of indigenous knowledge and local water policy negotiations, applying ecocritical analysis to the Waitanga Tribunal's processes and reports from 1975 to the present. Frequently, postcolonial studies in general—and postcolonial ecocriticism more specifically—consists of a litany of degradation and suffering, underscored by the impotence of native peoples and others who would resist forces of exploitation. The story of the Waitanga Tribunal, by contrast, is a unique story of empowerment, despite the fact that Maori people represent only 15 percent of New Zealand's population. Although initially rejected by the government, the community persisted and a dozen years later succeeded in agreeing upon a framework for settling the tribe's river claim in 2012. This contemporary narrative serves as a bright counterpoint to the all-too-typical narratives of bleakness and frustration.

Not far from New Zealand, though, nuclear industrialization is occurring in various Pacific Island nations and represents a considerable threat to both the social and the ecological fabric of the region. Nuclear weapons testing in French Polynesia and other areas of Oceania is well-known and dates back to World War II. It is less commonly known that such testing continued actively through the 1990s, and the ecological damage persists to this day and remains a focus of contention between the local inhabitants and the European (colonial) military establishments that have used the region as a de facto *terra nullius*, available for atmospheric weapons experimentation. Another aspect of this story that few people outside of the region know about is the way that local writers have developed a discourse of resistance. Dina El Dessouky, in her chapter, focuses on the work of Hawai'ian-born poet Haunani-Kay Trask and Marshallese poet Kathy Jetnil-Kijiner, which humanizes the "ocean people" who've lived through the era of nuclear testing. Readers of El Dessouky's study will be reminded of Tsutomu Takahashi's study of Minamata Disease and Inna Sukhenko's examination of Chernobyl literature in *Ecoambiguity, Community, and Development*—and, of course, other familiar stories of military testing and human health effects, such as Terry Tempest Williams's *Refuge.*

By this point, having considered a wide range of chapters written by both local authors and by scholars who live outside of the places represented in their studies, readers may be wondering why it matters who speaks for whom, who speaks for nature, and might the place itself ever, in some sense, have the potential to tell its own story. In a sense, this is what Benay Blend explores in her chapter about speaking with and for the natural world, a key theme in the work of such Native American authors as Linda Hogan (Chickasaw), Joy Harjo (Muskogee), and Louise Erdrich (Cree). The idea of situated, highly localized forms of knowledge and expression is a readily transferrable issue in ecocriticism, applicable to all of the regions (and more specific locales) represented in this book. But Native American writers have been among the most eloquent in articulating the concept of speaking with and for nature. Part of the significance of this topic is how it broadens to include the postcolonial conundrum of how to allow local human residents in previously colonized societies to speak forth with their unique perspectives of the world, perspectives which cannot, on some fundamental level, even be approximated by well-intentioned outsiders.

Remaining in the Americas, we turn next to Adrian Kane's study of Guatemalan and Costa Rican fiction—specifically, Rodrigo Rey Rosa's 1996 *Lo que soñó Sebastián* (What Sebastian Dreamt) and Anacristina Rossi's 1992 *La Loca de Gandoca* (The Madwoman from Gandoca). These literary texts abruptly challenge the idea of highly localized knowledge, existing independently of global markets and North-South political dynamics. In a sense, the tensions between Kane's essay and Blend's analysis of the local privileging that often emerges in Native American literature reveals an overarching tension, if not an outright contradiction, among many of the essays included in this project. In a sense, there is an idealistic, if somewhat unrealistic, hope that local wisdom and voices might exist in protected isolation from the depredations of global forces—and indeed, in Dawson's New Zealand case study, culturally specific ideas of nature (water) seem to have prevailed in convincing the government to grant protection to the local river. But all too often transcultural interactions have favored the outside cultures, which is the narrative told by such geographers as Jared Diamond in *Guns, Germs, and Steel: The Fates of Human Societies* (1999). In his essay, Kane departs from the more explicitly postcolonial framework of other essays in this book, relying on deep ecology to tease out the ecocentric thinking in *Lo que soñó Sebastián* that appreciates complex ecological systems in Guatemala which are imperiled by external economic forces, while using theories of addiction to explain Costa Rican acceptance of neoliberal economic systems, as depicted in *La loca de Gandoca*.

Continuing the eastward movement of the volume, our next contributor is James McElroy, who, though based in California, writes about Northern Ireland from a postcolonial ecocritical perspective. McElroy particularly emphasizes the notion of "provincializing discourse," which he adopts from ecocritic Lawrence Buell. The idea is that highly localized, or provincialized, forms of expression serve intrinsically as modes of resistance to homogenizing discourse that drowns the voices of the provinces within the voices of the colonizing culture. The first order of business in this essay, though, is to defend the notion that Ireland is a British colony and that Northern Ireland, more specifically, is "Ireland's real Global South." The author proceeds, by way of examining poetry from Northern Ireland, to offer a postcolonial ecocritical demonstration of the linkage between cultural and environmental damage caused by Britain's colonial presence. In particular, he works to present the nuanced local political and ecological contexts of poems by such writers as Tom Paulin, John Hewitt, Sam Robertson, Louis MacNeice, Michael Longley, Derek Mahon, John Montague, Alan Gillis, and Sinéad Morrissey—the lengthy list reinforcing the sense that this embattled territory has inspired not only violence but beautiful words. Sharae Deckard's abovementioned study of war ecology in Sri Lankan literature offers useful intersections with this chapter as well.

Reinforcing the relevance of postcolonial ecocritical analysis to Irish culture, Eóin Flannery next studies the twentieth-century Irish "Big House" novel as a genre especially suited to illuminating the colonial relationship between Britain and Ireland. He focuses specifically on Elizabeth Bowen's *The Last September* (1929), J. G. Farrell's *Troubles* (1970), and John Banville's *The Newton Letter* (1982), using ecocritical and ecopolitical lenses to read the Big House novels, which "represent the hubris and the decline of a semi-feudalist colonial system in Ireland." The colonial experience in Ireland, he argues, "was a fraction of what [Pablo] Mukherjee characterizes as 'a permanent state of war on the global environment,' conducted by 'colonialisms and imperialisms'." This chapter identifies the Irish Gothic, an aspect of the Big House novel, as a particular aesthetic device developed in this part of the world and well suited to depicting the decaying gentry.

Cameroonian ecocritic Augustine Nchoujie next draws our attention south to Africa, where he introduces emerging novelist Nsahlai Athanasius, whose 2008 work *The Buffalo Rider* vividly depicts the endangerment of animals in West Africa. Nchoujie describes Athanasius as "one of the voices crying out in the wilderness for Cameroon to protect her animals and plant species." In this novel, non-human animals are personified and de-objectified and are thus heralded as creatures that should be appreciated and protected, while human characters who hunt the buffalo are represented graphically as violent butchers. Set in the Mbar-Ngongmbaa-Kilum Mountains of Cameroon, this novel is deeply embedded in the country's landscape and culture. One of the functions of ecocriticism of the Global South is to highlight literary and other cultural texts that outsiders would be unlikely to know about and to explain the goals and subtle contextual aspects that non-local readers would likely miss without such guidance.

Just to the north of Cameroon is Nigeria, which boasts one of the largest and most vibrant literary communities in Africa. Senayon Olaoluwa has contributed a new ecocritical reading of one of Nigeria's iconic literary works, Chinua Achebe's *Things Fall Apart*. This novel has frequently been studied by postcolonial scholars and others, including ecocritics (such as Augustine Nchoujie). Olauluwa seeks to offer a different kind of ecocritical reading of Achebe's masterpiece, though, by broadening the focus from the traditional animist beliefs and practices of the characters to a more syncretic combination of traditional local beliefs and Christian ideologies. He thus pushes back against the somewhat reductionist view of the author's celebration of neo-animist belief systems as a route to ecological and cultural salvation. The on-the-ground reality of Achebe's novel and modern life in Nigeria, suggests Olaoluwa, is more complicated than this. As in many cultures throughout the world, the Nigerian experience is one of cultural pluralism, going beyond simple tensions between the old and the new, the native and the colonial. A global southern ecocritical practice must somehow appreciate this hybridity.

South African ecocritic Anthony Vital follows up with his examination of K. Sello Duiker's *Thirteen Cents* (2000), Zakes Mda's *The Heart of Redness* (2000), Ngũgĩ wa Thiong'o's *Devil on the Cross* (1980), and Doreen Baingana's *Tropical Fish* (2005), emphasizing not the stereotypical rural African landscapes and experiences that international readers might expect from that continent, but rather focusing on the urban African experience of "nature," which is the reality of many people there and throughout the world—he points out that more than half of the world's population are city-dwellers. This is true as well in South Africa, Kenya, and Uganda, where his authors locate their stories. Vital imagines an ecocriticism that can "explore, via the text, [. . .] the impact of urbanized life on our species and on the planet."

Even urban human societies, though, exist in a physical, geographical context, albeit a changing one. Iranian ecocritic Zahra Parsapoor examines Bozorg Alavi's "Gilemard" (1952), a well-known short story, which relates the experience of a man from Gilan, a forested region of northern Iran, who comes to the desert landscape of Baluchistan and suffers various hardships when interacting with the people there, especially the police. The characters in the story reflect the "personalities" of the regions they come from—there is a strong sense of geographical determinism here. Ultimately, as Parsapoor explains, Alavi's story pits the culture of village life against the cultural manifestations of the city. If Vital is correct in the previous article, it behooves us all to come to terms with urban life and its implications for future human life on the planet. Yet Parsapoor, in selecting the Alavi story for her close reading and in highlighting its critique of city culture (as a form of entropy), reveals that ecocriticism in her country continues to reinforce a pastoral ethos, one that idealizes small-town life and freedom from the political complexities of urban reality.

If nothing else, the dazzling array of international literary texts chosen to represent countries from around the world in this volume should impress readers with the idea that interesting, beautiful, and socially trenchant environmental expression is, essentially, a universal human cultural phenomenon. The North American and Western European "centrism" of ecocriticism and environmental artistic and journalistic expression is an illusion—or, rather, a delusion on the part of scholars based in North America and Western Europe. We ignore the voices and ideas of our colleagues elsewhere in the world at our own peril. The final essay in this book is Munazza Yaqoob's study of environmental consciousness on contemporary Anglophone Pakistani literature. Many readers of this book may never have heard of such authors as Uzma Aslam Khan and Mohsin Hamid, but we find in their work the ubiquitous concerns about environmental degradation and coming to terms with the reality of urban existence that are evident throughout world literature. What is particularly poignant in the literature and ecocritical commentary from the Global South, though, is the frequent critique of the impacts of global capitalism, a force largely transplanted from the Global North to the developing world, usually against the will of the human population of non-industrialized areas (and, of course, without any effort to consult with the natural environment).

If there is any doubt that we are all in this (this life, this predicament) together, we have only to consider the repeated motifs and arguments of this collection, accented by the edgy global southern focus on resistance to oppression instigated, to a great extent (but not always), by external forces. The field of ecocriticism has become a global phenomenon, but it is not practiced the same way everywhere. We hope this book, and our earlier collection *Ecoambiguity, Community, and Development*, will serve not only to celebrate the growth of this field, but as a means of drawing forth additional ecocritical voices from still silent communities.

WORKS CITED

Ashcroft, Bill, Gareth Griffiths, and Helen Tiffin. *The Empire Writes Back: Theory and Practice in Post-colonial Literatures*. London and New York: Routledge, 1989. Print.
Cronon, William, ed. *Uncommon Ground: Toward Reinventing Nature*. New York: Norton, 1996. Print.
DeLoughrey, Elizabeth, Renée K. Gosson, and George B. Handley, eds. *Caribbean Literature and the Environment: Between Nature and Culture*. Charlottesville: U of Virginia P, 2005. Print.
Diamond, Jared. *Guns, Germs, and Steel: The Fates of Human Societies*. New York: Norton, 1999. Print.
Nixon, Rob. "Environmentalism and Postcolonialism." *Postcolonial Studies and Beyond*. Ed. Ania Loomba and Suvir Kaul. Durham, NC: Duke UP, 2005. 233–51. Print.
———. *Slow Violence and the Environmentalism of the Poor*. Cambridge, MA: Harvard UP, 2011. Print.
Shiva, Vandana. *Ecology and the Politics of Survival*. New Delhi, India: Sage, 1991. Print.
Slovic, Scott, Swarnalatha Rangarajan, and Vidya Sarveswaran, eds. *Ecoambiguity, Community, and Development: Toward a Politicized Ecocriticism*. Lanham, MD: Lexington, 2014. Print.

Chapter One

The Environmentalism of
The Hungry Tide

Priya Kumar

In his groundbreaking study *Slow Violence and the Environmentalism of the Poor*, Rob Nixon draws attention to the reciprocal indifference and the broad silence that has characterized the relationship between postcolonial and environmental literary studies, and calls for a long overdue dialogue between the two fields.[1] Among the various schisms he identifies between postcolonialism and ecocriticism are their very different perspectives on questions of place and displacement: if postcolonial writing was largely preoccupied with displacement and discourses of cross-culturation and hybridity, environmental literary studies, until recently, tended to privilege the literature of place and a bioregionalist approach (236).[2] Although Nixon acknowledges "the emotional power generated by attachments to place," he goes to great lengths to illustrate the pitfalls of commitment to particular regional locales in American environmental writing, pitfalls which include hostility toward displaced people, jingoism, and "super-power parochialism" (242, 258). Nixon recognizes that there is nothing intrinsically good or bad about place attachment—it can instill a conservative environmental ethic or a more progressive inclusive one—but his emphasis is on pointing out the dangers of what he calls an "ethics of place environmentalism" in much American writing and criticism (241). While he also takes postcolonial literary critics to task for their belated attention to environmental concerns—which they tended to regard as elitist or sullied by green imperialism—he affirms the importance of postcolonialism in helping to expand environmental thinking beyond the dominant paradigms of purity, wilderness preservation, and Jeffersonian agrarianism (243). Accordingly, he calls for a (less exclusionary) "transnational ethics of place" that draws on the strengths of bioregionalism without entailing the kind of spatial amnesia toward non-American (or rather non-U.S.) geographies that bioregionalism has often implicitly rested upon, and that incorporates experiences of hybridity, displacement, and transnational memory (243).

If we are to respond to Nixon's call for a "radically creative alliance" between environmental and postcolonial literary studies, we certainly must heed the work of writers who enable us to "apprehend" experiences of displacement and slow violence that are spatially and historically remote from us, as he suggests, but we must also take seriously the cultural and ethical work that "place-responsive imaginative acts can perform" for marshaling environmental concern (Nixon 259; Buell, *Writing for an Endangered World* 64).[3] Nixon tends to undervalue the role that a powerful sense of place can play in mobilizing people against environmental degradation given his critique of the kind of blinkered place-sense we encounter in a certain tradition of American environmental writing. In contrast to Nixon, Lawrence Buell argues that the many abuses of platial attachment don't invalidate it. Underlining the significance of place for environmental resistance movements, Buell writes, "the more a site feels like a place, the more fervently it is so cherished, the greater the potential concern at its violation or even the possibility of its violation" (56).[4] Indeed, he suggests that there is no physical space that is not potentially a place and hence it is in the interests of the "planet, people, and other forms of life . . . for 'space' to be converted—or reconverted—into 'place'" (78). At the same time, he is careful to qualify that place-sense should not be limited to what we think of as home. A progressive sense of place must be understood as entailing "many different patches besides just home" and place will become truly meaningful, particularly for environmental thinking, only when we are able to think of place at the scale of the planet (77).

This chapter provides a close reading of Amitav Ghosh's novel *The Hungry Tide* in light of these broader questions about displacement and attachment to place raised by Nixon and Buell. Set in a rural and marginalized space of the Global South, the Sundarbans islands on the borderlands between India/Bangladesh, *The Hungry Tide* is a powerful postcolonial environmental fiction that invites us to apprehend stories of displaced people who are seen as disposable beings (Nixon), and simultaneously evinces a profound commitment to the ecologically rich but vulnerable deltaic region of the Sundarbans (Buell). It demonstrates that postcolonial and ecocritical concerns do not necessarily have to be in conflict with each other. Indeed, I would argue we must question the very binary that Nixon sets up between displacement as a postcolonial concern and an "ethics of place" as an ecocritical concern (238). Displacement certainly has been a major thematic among some prominent metropolitan postcolonial critics and cosmopolitan writers, but place making and the affirmation of place against place-eroding historical forces such as war and ethnic/religious violence also continue to be important concerns in postcolonial fiction, particularly in narratives of displacement and dispossession. For example, evoking and creating a sense of place has long been a preoccupation of much South Asian literature in English as well as in Indian languages. Qurratulain Hyder's Sindh, East Bengal, and Lucknow, Attia Hosain's Lucknow, Salman Rushdie's Bombay, Kamila Shamsie's Karachi, and Siddhartha Deb's (unnamed) Shillong are some of the literary constructions of place that come immediately to mind. Ghosh's *The Hungry Tide* belongs to this genre of place-making narratives; however, in contrast to much post-1980s Indian-English fiction, which has tended to evoke and construct urban metropolitan places, *The Hungry Tide* focuses on the rural and peripheral locale of the Sundarbans.

This chapter draws attention to two important narrative strands of the novel: in the first section, I read *The Hungry Tide* as an important environmental justice narrative that places the suffering of dispossessed and marginal human beings in the Global South at the forefront of its ethical and aesthetic concerns. I examine Ghosh's representation of the Sundarbans as a subaltern region characterized by human suffering and catastrophe; in particular, I address the novel's compelling critique of wilderness and wildlife conservation projects such as Project Tiger that are undertaken without any regard for their human cost. Ghosh's novel makes clear that environmentalism does not have to be antihuman, as the very spectrum of what counts as environmentalism has expanded with the increasing visibility and credibility of environmental justice movements.[5] However, *The Hungry Tide* is also a powerful ecocentric fiction that speaks "to the state and fate of the earth and its nonhuman creatures" by means of its imaginative account of the tide country (Buell, *Future* 127). In the second part of the chapter, I examine Ghosh's construction of the Sundarbans as a fascinating place that is home to a dazzling variety of flora and fauna, but one that is deeply vulnerable to the whims and particularities of the daily tides, the ever-present danger of cyclones and storms and, above all, to human carelessness and arrogance. I argue that the tide country serves as a compelling locale in the novel that allows Ghosh to bring together his environmental justice concerns about the horrific treatment meted out to indigent East Bengali refugees in the name of tiger conservation along with posing important questions about the fragility and vulnerability of the nonhuman world of the Sundarbans. By evoking and creating the Sundarbans as a place, he seeks to marshal concern amongst his cosmopolitan readership about this endangered region and its indigent inhabitants.

Told largely from the perspective of two outsiders to the Sundarbans—Kanai Dutt, an affluent upper middle-class Bengali man who runs a translation and interpretation bureau in Delhi, and Piya Roy, an Indian-American of Bengali origin, who is a cetologist researching the once teeming population of dolphins in the tide country—the novel alternates between Kanai and Piya's narratives and their transformative experiences in the tide country, or the *bhatir desh*, as it is known to its inhabitants. Kanai is visiting the fictional island of Lusibari at the behest of his aunt Nilima to collect and read a diary that has been left for him by Nilima's late husband, Nirmal, a former leftist whose withdrawal from political activism led them to settle in the Sundarbans many years ago. Interweaving Kanai's reading of Nirmal's diary with Piya's search for the Irrawaddy dolphin, Ghosh's novel portrays how the affective encounters of its two central characters with the islanders and the natural world of the Sundarbans lead to an uncanny awakening of self in both (although the transformation is much more marked in the case of Kanai). By the end of their individual narrative trajectories, Piya and Kanai are able to establish a sense of place-connectedness to the tide country and its people in ways that resonate with both Nixon's notion of a transnational ethics of place and Buell's understanding of place-sense as containing many different patches besides home.

DISPOSABLE PEOPLE IN *THE HUNGRY TIDE*

Ghosh begins his narrative by drawing attention to the geographic marginality of the Sundarbans islands as a borderland region stretching for almost two hundred miles between India and Bangladesh, referring to the region as "the trailing threads of India's fabric, the ragged fringe of her sari" (6). This image conjures the Sundarbans literally as India's periphery or "fringe"—a far-flung place even within the Global South. The novel highlights Nirmal and Nilima's initial reaction to the "strangeness" of the tide country, given the marginalization of the islands in the national imagination: "the realities of the tide country were of a strangeness beyond reckoning" (66–67). The deprivation of the tide country reminds them of the terrible famine that had devastated Bengal in 1942, except that hunger and catastrophe were part of the everyday reality of the islanders. The novel disrupts the dominant *bhadralok* image of the Sundarbans as a beautiful garden devoid of human beings—based on a romanticized urban vision of nature and wildlife—by confounding its urban, bourgeois characters with the alterity of the tide country.[6] Anthropologist Annu Jalais, in her essay "Whose World Heritage Site?," asks us to consider the implications of the notion that "humans do not or should not fit in the Sundarbans" (2). As she points out, the Sundarbans are not just forest or *jongol*; they are also an inhabited region or *abadh*, something that tends to be elided in most media and scientific representations of the Sundarbans. In the novel, the tide country serves as a strange and unfamiliar landscape where elite, urban characters like Nirmal, Nilima, Kanai, and Piya find themselves in a version of the quest narrative, but it does not merely function as a backdrop—or a "prop" (to invoke Chinua Achebe's critique of the representation of Africa in Joseph Conrad's *Heart of Darkness*)—for their inner journeys. Rather, the social history and the natural environment of the Sundarbans are at the forefront of Ghosh's novelistic concerns.

Kanai's encounter with the inhabitants and the social history of the Sundarbans is mediated largely through Nirmal's notebook, which is a personal account of Nirmal's own life-altering experience in the tide country at the time of what came to be known as the Morichjhapi massacre. Written with a sense of urgency over a period of a day-and-a-half, Nirmal's diary is in the form of a testimony addressed to Kanai in which Nirmal seeks to bear witness to the massacre and brutal eviction of East Bengali refugees from the Sundarbans island of Morichjhapi.

Like many Partition refugees, refugees from East Pakistan/Bangladesh saw themselves as victims of history and believed that it was their legitimate right to seek refuge in the Indian part of Bengal.[7] However, while earlier migrants from East Pakistan were accommodated in Calcutta and its environs, later migrants found it difficult to claim a place in the state despite their shared Bengali "Hindu" identity, which had to do with differences in the class and caste composition of the two groups.[8] Many of the later arrivals belonged to low-caste groups engaged in occupations such as fishing, boating, paddy cultivation, and carpentry. The government of West Bengal viewed these later arrivals as a joint burden to be shared with the central government and the neighboring states. Likewise, the West Bengal elite saw the refugees from East Pakistan as an economic burden that would drain the already meager resources of the state. Accordingly, the poorest of these refugees were forced to move to various inhospitable and infertile areas in central India such as Dandakaranya and Mana in the late 1950s and throughout the 1960s under pressure from the provincial government. In 1978, when the Left returned to power in Bengal, nearly 120,000 refugees sold off their belongings and returned to West Bengal in the hope that the new government—that had taken up their cause in the past—would rehabilitate them in the state. About 25,000 of these refugees sailed to the island of Morichjhapi in the Sundarbans and decided to settle there. However, the new Left government intercepted many of the refugees on their way to Bengal and sent them back to the camp at Dandakaranya. As for those who had settled in Morichjhapi, the government declared it to be an illegal encroachment on forest land in an area that was reserved for endangered tigers and took it upon itself to evict the settlers from the island. Many refugees died or were injured as a result of police action, while others died from starvation or eating food unfit for human consumption.

The Hungry Tide chronicles the story of these thrice-displaced East Bengali refugees who were never incorporated into the new national order of

India, despite the official homecoming narrative of the Indian state toward "Hindu" and Sikh refugees.[9] The novel attests to their collective experience of displacement and eviction through Nirmal's diary. Witnessing the horrific events at Morichjhapi brings to life Nirmal's creative energies and impels him to write after a long hiatus of thirty years. Nirmal is very self-conscious of his role as a witness and about the responsibility of testifying to events that will soon be forgotten. "[H]ow skillful," he writes, "the tide country is in silting over its past" (59). Throughout his notebook, Nirmal constantly evokes the poet Rainer Maria Rilke, and each section ends with lines from Rilke's *Duino Elegies*. Toward the end of his poignant account, Nirmal recalls Rilke's injunction: "this is the time for what can be said. Here is its country. Speak and testify . . ." (227). As Jacques Derrida notes, "In principle, to testify—not being a witness, but testifying, attesting, 'bearing witness'—is always to render public. The value of publicity, that is, of broad daylight (phenomenality, openness, popularity, res publica, and politics) seems associated in some essential way with that of testimony" (30).[10] This mandate to testify, to bring to light forgotten histories and dispossessed marginal historical subjects, serves as the imperative for *The Hungry Tide*, and indeed for much of Ghosh's fiction.

At the center of Nirmal's notebook is the story of Kusum, a young woman from the Sundarbans who had made her way to the island of Morichjhapi along with the refugees from the Dandakaranya camp. Nirmal sees Kusum as his muse and identifies her with the excitement of revolution and poetry in contrast to his wife Nilima whom he associates with prose and mundane change.[11] Kusum meets the refugees in the mining town of Dhanbad as they are on their way to the Sundarbans. The refugees tell Kusum that they are "tide country people" from the edge of the Sundarbans (136). They crossed the border when the war broke out (presumably 1971, but this isn't specified) because their village was burnt down and *"there was nowhere else to go"* (136, my emphasis). Ghosh tries to find a language to describe the topophilia of these nowhere people as they talk of how they were taken by the police to the settlement camp in central India and how they were unable to settle in that "dry emptiness" because "the rivers ran in our heads, the tides were in our blood" (137). A former inhabitant of the Sundarbans herself, Kusum joins the refugees and marches with them to Morichjhapi where she meets Nirmal after many years as he seeks shelter from a tide-country storm.

Through Nirmal's diary, Ghosh goes to great lengths to highlight the resilience and the courage of the refugees in contrast to the dominant construction of refugees as passive victims. He describes the heroic efforts of the refugees to make a place out of Morichjhapi in a matter of a few weeks. They had built the embankment, which was crucial for their survival on the island; paths had been cleared; plots of land had been enclosed with fences (141). What is more, they had set up their own government and taken a census to get a sense of how many people there were on the island. Nirmal is astounded by these developments—it was as though "an entire civilization had sprouted suddenly in the mud" (141). Ghosh points to the very different representational conventions that refugees are subject to—those herds for which UN agencies are created—as compared to the elevated individual exilic figure or writer.[12] Nirmal realizes that he had imbibed the lesson of such conventions when he can only think of the "rifugi" as part of an untidy mass in need of aid and charity as opposed to the intrepid settlers he encounters on the island (141).

Ghosh goes on to recount the dreadful events of Morichjhapi in May 1979 and the resistance of the refugees to the state's attempts to eject them. Nirmal learns that all movement in and out of the island has been banned under the Forest Preservation Act and that it had been surrounded by police boats; an economic blockade has been put into effect and the settlers have been forced to eat grass due to a shortage of food. When Nirmal attempts to go to the island with a group of schoolmasters, he witnesses a boatload of defiant refugee people trying to resist the police order as they shout in unison: "Who are we? We are the dispossessed . . ." (211). Ghosh's use of the Bengali word *bastuhara* to express the grief and defiance of the refugees is significant here. *Bastuhara* is, literally, one without objects or possessions (those which make a home). Rather than fashioning themselves as a diasporic people with links to an ancestral homeland, these East Bengali refugees—who have been displaced by the multiple partitions of the subcontinent—see themselves as nowhere people who have been rendered homeless by the very state that claims to represent them as "Hindu" refugees.

In one of the most powerful episodes in the novel, aptly titled "Crimes," Ghosh highlights how Morichjhapi has been a crime against humanity, one that has conveniently been forgotten.[13] A weak and emaciated Kusum, in a stirring indictment of environmental conservation policies that are undertaken at the expense of poor and marginal human beings who live in proximity to tigers, tells Nirmal that the worst part of the police blockade was not so much the hunger or the deprivation, but the recognition that they were designated as disposable beings whose lives were of no value to the state or to elite Bengali society: "Who are these people, I wondered, who love animals so much that they are willing to kill us for them?" (216). Kusum is referring to India's flagship conservation project, Project Tiger, which was launched in 1973 with substantial impetus and funding from the World Wildlife Fund and the International Union for the Conservation of Nature and Natural Resources as part of an attempt to transplant the American system of national parks to India. The Project Tiger reserves are part of a system of 75 national parks and 425 wildlife sanctuaries that cover about 4 percent of India's surface area.[14] Project Tiger propelled the Sundarbans area into fame as the largest remaining natural habitat of the Royal Bengal tiger. However, as

Ramachandra Guha argues in his trenchant critique of wilderness preservation, "because India is a long-settled and densely populated country in which
agrarian populations have a finely balanced relationship with nature, the
setting aside of wilderness areas has resulted in a direct transfer of resources
from the poor to the rich. . . . The designation of tiger reserves was made
possible only by the physical displacement of existing villages and their
inhabitants," creating a new category of "ecological refugees" in the process
(Guha 95, 107). As refugees from East Pakistan/Bangladesh, the Morichjhapi
settlers were already thrice displaced; Project Tiger made (ecological) refugees of them yet again.

Kusum articulates her resentment that people like her, the disinherited
disposable beings evoked by Rilke, are viewed as a form of "human waste,"
superfluous, irrelevant, and redundant populations that are seen as burdening
the earth, who must be properly disposed of, much like the world tries to get
rid of its unwanted garbage. Zygmunt Bauman underlines how refugees are
designated as a form of human waste because they are viewed as having no
useful role to play in the land of their arrival, with no prospects of being
incorporated into the new social body (77). Once they are assigned to the
category of waste, their chances of being "recycled into legitimate and acknowledged members of human society are, to say the least, dim and infinite-
ly remote. All measures have to be taken to assure the permanence of their
exclusion" (78). Thus denied even the barest minimal rights of survival,
refugees must be sent to "dumping sites" or to distant places to ensure that
they can be held at a safe distance—or quarantined—from those who are
already inside (77–78). Bauman's analysis is especially pertinent for understanding the situation of Dalit and other lower caste refugees who migrated
to West Bengal later. If we understand the Dandakaranya camp as a "dumping site" to which these poorer migrants from East Bengal were dispatched
to, then we can appreciate why the refugees' act of returning to West Bengal
and settling on the island of Morichjhapi in direct defiance of the directive of
the West Bengal government is seen as immensely threatening to the authority of the state government. Hence, they must be expelled. Kusum understands that refugees and dispossessed people like her matter less than the
Royal Bengal tiger, who has been declared the national animal of both India
and Bangladesh. Significantly, Ghosh also uses Kusum to chastise the affluent urban readers of this novel, who may often support wildlife conservation
unthinkingly, for our complicity in the human rights' abuses of Morichjhapi.

As the novel unfolds, it traces the affective impact of Nirmal's diary on
Kanai who, as a young boy had described the Sundarbans as "rat-eaten islands," but gradually comes to realize how elite Indians like him and even
conservation biologists like Piya are implicated in the ongoing dispossession
of people who live in the Sundarbans. In an absorbing episode, Ghosh describes how a tiger, which has been accidentally trapped inside a livestock

pen, is being killed by the islanders. Piya is horrified by what she views as an archaic scene of mob violence and cannot understand how the islanders would want to take revenge on an animal. She can only make sense of this scene as "the horror," but Kanai demurs and reminds her that both he and she, as representatives of the outside world, are complicit in the horror—she, as part of those who support efforts to protect the wildlife of the Sundarbans without any regard of the human cost, and he, because Indians of his class have chosen to ignore or gloss over dispossessed beings in order to appeal to Western patrons (249). He tells her that people are killed by tigers almost every week in the Sundarbans but no one pays any heed to such people because they are seen as disposable people who are "too poor to matter." He asks her: *"Isn't that a horror too—that we can feel the sufferings of an animal, but not of human beings?"* (248, my emphasis). By highlighting the terrible cost of privileging animals over people—especially poor and disenfranchised people—the novel provides a powerful commentary on species and wilderness preservation projects like Project Tiger that are undertaken without any care for the marginal human beings who live in the vicinity of wildlife sanctuaries and national parks and who are often dispossessed in the name of such projects.

Ghosh's environmental justice concerns are very evident in this strand of the narrative. The environmental justice movement accuses the global environmental establishment—mainstream Western organizations such as the Wilderness Society and the World Wildlife Federation—of undue concern with endangered species at the expense of marginal and oppressed humans. As Buell points out, "preservationism in the Thoreau-Muir tradition, was to become the operating philosophy of the elite environmental organizations against which environmental justice activism has often pitted itself" (*Writing for an Endangered World* 38). The environmental justice movement differs in several core values from the environmental establishment: concerned with the impact of environmental degradation on poor and marginalized populations, environmental justice activists and theorists view environmentalism as an "instrument of social justice" and interhuman equity as opposed to caring for nature as a good in itself. Increasingly led by non-elites, it is more explicitly "anthropocentric" in its ethical concerns than the ecocentrism or biocentrism of the dominant preservation ethos (38).[15] Buell goes on to qualify that this is by no means to suggest that environmental justice advocates don't care for the earth; however, they reject "an ethic of concern *only* for nature preservation" at the cost of people, especially poor and disempowered groups (279, 281).[16]

The environmental justice movement's critique of the homogeneity of elite environmentalism and academic environmental studies has had an immense impact on the field of literature and environment studies such that scholars now distinguish between first-wave and second-wave ecocriticism in recognition of the expansion of critical horizons within the field. If the hallmark of first-wave ecocriticism was a strong ecocentrism—"the view in environmental ethics that the interest of the ecosphere must override that of the interest of individual species," especially humans (Buell, *Future* 137)— and a partiality for genres like nature-writing and wilderness fiction—then second-wave ecocriticism is distinguished by a "sociocentric" ethics and aesthetics that has been strongly influenced by the environmental justice movement's engagement with issues of environmental welfare and equity to socially marginalized people (Buell, Heise, and Thornber 433).[17] How do we reconcile these cleavages within environmental criticism and writing?[18]

Buell maintains that the soundest positions will be those that come closest to "speaking *both* to humanity's most essential needs and to the state and fate of the earth and its nonhuman creatures independent of those needs, as well to the balancing if not also the reconciliation of the two" (*Future* 127).[19] Although *The Hungry Tide* has most often has been read in light of its imaginative attempt to bear witness to the history of Morichjhapi and its critique of wilderness and wildlife preservation, I contend that the novel is also a compelling ecocentric fiction.[20] While the novel is clearly preoccupied with questions of justice for people like Kusum and the Dandakaranya refugees who have been evicted in the name of preserving and protecting tigers, it also manifests a profound concern about the fragility and the environmental future of the natural world of the Sundarbans. To do justice to the novel's complexity, we must also attend to its ecocentric concerns.

THE ECOCENTRISM OF *THE HUNGRY TIDE*

Ecocentrism encapsulates a range of philosophical positions, but what is common to the various strains of ecocentric thinking is the belief that human identity must be understood in terms of its imbrication and interdependence with the physical environment and/or nonhuman entities. Ecocentrism maintains that the "world is an intrinsically dynamic interconnected web of relations" with "no absolute dividing lines between the living and the nonliving, the animate and the inanimate" (Eckersley, cited in Buell, *Future* 137). This kind of philosophical orientation informs the very texture and the form of *The Hungry Tide*.

Throughout, the novel emphasizes the ephemerality of the physical environment of the Sundarbans and its hostility to human settlement: "The tides reach as far as two hundred miles inland and every day thousands of acres of forest disappear underwater, only to reemerge hours later" (6–7). Mangrove forests gestate overnight, we are told, and visibility is extremely limited because the forest is very dense. Every year, dozens of people are killed by tigers, snakes, and crocodiles. Ghosh highlights how the particularities of the environment make it an everyday struggle for human beings to make a place out of this inhospitable terrain.[21] The twice-daily tides, the constant threat of storms and cyclones, the erosion of the embankment, and the dangers posed by wildlife make human life on the islands the result of remarkable resilience and creativity. At the same time, Ghosh also dramatizes the threats posed by human activities and human settlement to the aquatic and terrestrial life forms of the tide country. The Sundarbans is evoked as a fragile and vulnerable ecosystem that is in perpetual danger of being eroded. As Supriya Chaudhury insightfully notes, "the imminence of disaster, whether natural or human, covers the world of this novel with a kind of film by which 'precious objects appear doomed and irrevocable in the very moment of their perception,' as Freud said in his essay 'On Transience'."

All through, the novel draws attention to the natural riches of the islands and revels in their abundant flora and fauna. Ghosh includes many long descriptive passages and minute scientific details about the tides, the porous and permeable environment of the tide country, the mangrove forests, as well as the many different animal species that inhabit the Sundarbans such as dolphins, crabs, tigers, and crocodiles. We are told that there are more species of fish in the Sundarbans than in the entire continent of Europe: the waters of river and sea interpenetrate unevenly in this deltaic region creating "hundreds of ecological niches" or "microenvironments" that are filled with their own flora and fauna (105). Like many environmental writers such as Michiko Ishimure, Don DeLillo, and Christa Wolf, Ghosh incorporates scientific facts and details into his narrative in order to shore up his ecocentric preoccupations. This interweaving of scientific minutiae with story-telling has been a hallmark of much environmental writing—"a kind of prose and film that sits at the intersection of narrative and science, blending the endeavor to convey a scientific perspective on environmental crisis with the impulse to tell large- and small-scale stories about humans' interaction with nature" (Buell, Heise, and Thornber 423).

For example, Ghosh's interest in conservation biology and his ecological concerns about accelerated species loss due to ruined habitats are especially evident in his detailed account of the *Orcaella brevirostris*, the Irrawaddy dolphin. He expresses his ecocentric commitments primarily through Piya and her anxiety about the dwindling cetacean population of the Sundarbans, which has been disappearing from its established habitat over the years because of human activities and carelessness. Ghosh dramatizes this utter disregard toward nonhuman others in an episode where Piya and Fokir come across the carcass of a newborn Irrawaddy dolphin. Piya realizes that the calf had probably been hit by a fast-moving motorboat, probably an official boat, manned by the very agents of the state who are supposed to implement conservation policies on the ground. Piya believes her work on the Orcaella could be crucial toward the preservation and sustenance of this endangered species. She counters Kanai's powerful and stirring indictment of global conservation efforts by warning him of the dangers of speciesism or what one could call an incontrovertible anthropocentricism—the arrogance of the human belief that no other species matters except us. She tells him:

> *Just suppose we crossed that imaginary line that prevents us from deciding*
> *that no other species matters except ourselves.* What'll be left then? Aren't we
> alone enough in the universe? (249, my emphasis)

Here Piya emphasizes the interdependence of disposable humans and nonhumans and suggests that more anthropocentrically inclined environmental justice commitments can coexist with ecocentric ideals. They don't necessarily have to be exclusionary. It is not insignificant that the chapter in which this dialogue takes place is titled "Interrogations"—in the plural—indicating how both Kanai and Piya's perspectives, at different points, are held up to critical scrutiny in the text.

The figure of Nirmal also serves to articulate many of Ghosh's concerns about the fragility of the natural environment of the Sundarbans. Expressing an ecocentric perspective, he believes that "everything which existed was interconnected: the trees, the sky, the weather, people, poetry, sciences, nature" (233). However, he also articulates his apprehensions about the gradual depletion of the once plentiful flora and fauna of the tide country. He remembers how he used to see the sky darken with flights of birds when he first came to Lusibari, and how the mudbanks used to swarm with crabs. But now the birds and fish are disappearing. Nirmal is acutely aware that it wouldn't take much to submerge the tide country given the particularities of the natural environment—where the land emerges only at low tide—and where "transformation is the rule of life" (186). For example, the novel points to the dangers of indiscriminate deforestation through Nirmal's account of the rise and fall of the town of Canning when the British decide to build a new port there in the nineteenth century. The mangroves, we are told, were Bengal's protection against the encroaching sea. But, the colonial administrators pay no heed to the warnings of Henry Piddington, a minor official who used to study storms and kept warning the planners and surveyors of the dangers of building a town so deep in the tide country by destroying the forest. Eventually, the "dangerously exposed" port is destroyed by a relatively minor storm and is abandoned in 1871. (This yesteryear storm anticipates the storm at the end of the narrative that leads to Fokir's death and indiscriminate destruction in the islands.)

Above all, the novel manifests its ecocentric ethical orientation by shining a light on indigenous forms of imagination and the role they can play in supporting the natural environment of the Sundarbans. As an academic field, ecocriticism has long been drawn to indigenous art and imagination. As

Buell et al. point out, within ecocriticism's broad interest in indigenous environmental works several concerns stand out: indigenous place-based stories and myths and what they tell us about long-term collective attachment to specific locales; "the nondualistic recognition within 'native' peoples collective imagination of nonhuman entities as fellow beings, whether at a sensory or spiritual level or both"; the ways in which indigenous art and thought attest to multiple forms of environmental injustice and resistance (429). I examine the novel's recurrent invocation of one such place-based story that is indigenous to the Sundarbans and its account of how humans should relate to their natural environment and to nonhuman others.

Ghosh deploys the legend of Bon Bibi—the goddess of the forest, who is worshipped by many of the Sundarbans islanders, especially those who work in the forest—to consider the crucial ethical question of how should one live with nonhuman others, the other as alter-species, in contrast to a philosophical tradition that, since Descartes, seeks to elevate man over and above all living creatures. [22] He describes it tellingly as "the story that gave this land its life" (292). As Ghosh points out in his essay "Wild Fictions," this story is almost unknown outside the Sundarbans, but it saturates the experience of those who live there such that no islander ever ventures into the forest without calling upon Bon Bibi for her protection. Given the author's fascination with syncretic religio-cultural practices in earlier works such as *In an Antique Land* and *The Shadow Lines*, it is not surprising that he finds the story of Bon Bibi so appealing. When Piya and Nirmal, on different occasions, observe Fokir and Kusum praying to Bon Bibi, they are surprised by the use of Arabic invocations in what appears and sounds like a typical Hindu puja. Nirmal discovers that, according to the legend, Bon Bibi was born in the city of Medina in Arabia, and that the printed version of the Bon Bibi story was attributed to a Muslim man named Abdur-Rahim (205). [23] Yet, it is not just the Muslim provenance of Bon Bibi and the fact that she is worshipped by many lower-caste Hindus in the Sundarbans that fascinates Ghosh, but also what the legend has to say about respecting and revering the space of nonhuman others. Bon Bibi, we are told, came to the Sundarbans along with her brother Shah Jongoli to make the tide country "fit for human habitation," which at the time was ruled by a demon called Dokhin Rai, who often took on the guise of a tiger and attacked humans (86). [24] After Bon Bibi defeats Dokhin Rai, she decides to divide up the tide country between Dokhin Rai and human beings. She determines that one part of the tide-country would remain a wilderness to be allotted to Dokhin Rai, while the other should be made safe for human settlement. In this way, the Bon Bibi story establishes "the forest as a kind of commons to which all have equal access" (Jalais, *Forest of Tigers* 72).

As Ghosh's narrator comments, thus "order was brought to the land of eighteen tides, with its two halves, the wild and the sown, being held in

careful balance" (86). The key word here for Ghosh is "balance." Bon Bibi is
the deity who maintains a balance between the forest and the sown and the
upholding of this balance rests upon the islanders' belief in her. Ghosh elaborates on his understanding of the significance of the legend in "Wild Fictions": "the Bon Bibi legend uses the power of fiction to create and define a
relationship between human beings and the natural world. . . . It is not of
course unique in its vision of the relationship between human beings and the
natural world: similar conceptions of balance, reverence, and the limitation
of greed are to be found in many other places" (59). Although the term
"nature" or its equivalent does not appear in the legend, Ghosh believes that
its consciousness is ever present and that the narrative of Bon Bibi is profoundly informed by "ecological concerns" (59). Still today, the islanders
believe that the *jongol*—the uninhabited islands to the South—is the domain
of Dokhin Rai (and by extension of nonhumans) and that they must enter the
forest only when they really need to go there.[25] The forest is for the poor who
have no intention of taking more than they need; it must not be approached as
a site of resource extraction. And when the islanders do go into the *jongol*
they must not leave any trace of their presence—for example, by urinating,
spitting, or defecating—or else they will invite the wrath of the tiger-demon.

This attitude of respecting the nonhuman other's right to the forest is very
evident in the worldview of the subaltern characters in the novel including
Kusum, Fokir, and Horen. When Nirmal accompanies Kusum and Horen to
the island of Garjantola, an island deep in the jungle, where Kusum's father
had built a shrine to Bon Bibi, he is bemused to see all three of them bowing
in obeisance as they cross a certain part of the river. He is told that this is the
point at which they "*had crossed the line Bon Bibi had drawn to divide the
tide country*" (185, my emphasis). Kusum and the others are scared because
they have gone across the border that separates the realm of human beings
from the domain of Dokhin Rai. Nirmal realizes with a sense of shock that
"this *chimerical* line" was as tangible as a barbed wire fence may be to him
(186, my emphasis). This episode recalls the same invisible boundary that
Piya invokes when she warns Kanai of the dangers of a rampant anthropocentric worldview that seeks to subdue and conquer all other living beings
(249). In effect, Piya is saying that we must not cross this line. The relay of
shared images between the two passages separated by over fifty pages is not
coincidental. Both emphasize the notion of a self-imposed ethical border that
only holds meaning so long as we believe in it—the difference is that one
evokes a secular tradition (environmentalism, the extension of "human"
rights) and the other emerges from a place-based religious tradition (the Bon
Bibi story).[26]

Ghosh comments on the relative success of Project Tiger in the Sundarbans as compared to many of the project's "showcase reserves" where the
tiger population dwindled very substantially despite the outlay of huge

amounts of money and the displacement of many people. He suggests the fact that tigers held their own in the Sundarbans—despite the many human mortalities—has probably more to do with the power of the Bon Bibi story than the efficacy of any governmental project. In "Wild Fictions," he writes, "while it is by no means the case that indigenous people are always good custodians of the environment, neither is it true that their practices are always destructive. Today, it is widely accepted that many such groups have indeed played an important part in the preservation and maintenance of forests and ecosystems" 60).[27] Ghosh's novel implies that these indigenous narratives and forms of knowledge have been marginalized by the state and global conservation discourses, and that their energies must be harnessed to support contemporary environmental efforts.[28]

Jalais provides a fascinating coda to the section of the Bon Bibi story that focuses on Dukhe's tale. Dukhe is a poor young boy who is attacked by Dokhin Rai in his avatar as a tiger after Dukhe is betrayed by his greedy Uncle Dhona. When Dukhe calls upon Bon Bibi to save him, she immediately comes to his rescue and instructs her brother Shah Jongoli to teach the tiger-demon a lesson, who runs for his life. Ghosh's account of this part of the legend ends here, but Jalais goes on to tell us that Dokhin Rai runs to his friend, the Gazi, a pir (an Islamic holy man), who advises him to ask for Bon Bibi's forgiveness and to address her as "mother." Subsequently, Bon Bibi accepts Dokhin Rai as her "son" but not before Dokin Rai also has his say. He cautions Bon Bibi that if humans were to be given a free reign of the forest, there won't be any forest left to speak of. And so, Bon Bibi, as the just goddess of the forest, makes all three of them—Dukhe, Dokhin Rai, and the Gazi—promise that they would treat each other as "brothers" to ensure that humans and nonhumans will not be a threat to each other (*Forest of Tigers* 72). The imaginative hold of the Bon Bibi story thus enables the Sundarbans islanders to envision nonhuman others as fellow beings and to envision a form of coexistence with them that is quite different from the idea of species preservation. As Jalais reminds us, the legend positions the animals of the forest as Bon Bibi's "elected kin": because Bon Bibi suckled a deer as a child, she is believed to consider the deer as her mother; since she has adopted Dokhin Rai—the half sage-half tiger—as her son, she is considered to be related to tigers as well (85). The story thus ties tigers and forest fishers in a web of interdependence through the same symbolic mother; moreover, both depend on the forest to survive and both share the same harsh environment of the Sundarbans, which makes irascible beings of them (74).

In the novel, Ghosh tends to highlight the legend's notions of balance and respect for the dividing line between the forest and the sown, as well as the importance of regulating human greed, but he does not focus as much on the aspect of the story that imagines tigers as fellow beings ("brothers") perhaps because his environmental justice concerns take precedence. For example, in the chapter titled "the Megha," Nilima draws attention to the number of people killed by tigers in the Sundarbans every year and talks about the absurdity of forest departmental programs that sought to provide fresh water to tigers in an area where human beings periodically go thirsty. The impact of this account on Kanai is made very evident later in the novel when he describes these tiger-related casualties as "genocide" to Piya—thus mistakenly endowing the tiger with human intentionality (248). And, of course, as we have seen, Ghosh's recounting of the Morichjhapi episode serves to provide a critique of tiger conservation policies that are implemented at the cost of dispossessed humans.

Yet, interestingly, Ghosh *is* able to imagine the dolphins as related beings perhaps because dolphins do not pose any risks to humans: thus Kusum and Fokir refer to the dolphins as "Bon Bibi's messengers" and believe that dolphins bring Bon Bibi "news of the rivers and khals" and they help fishermen like Fokir find fish (254). Humans and dolphins are thus tied together through Bon Bibi in Ghosh's imaginative rendition of the legend, *but the novel stops short of envisioning tigers as fellow entities* in the ways that Jalais's anthropological account highlights even as Ghosh devotes a lot of narrative space to the figure of the tiger as compared to other nonhuman entities. I agree with Kaur's assessment that the tiger becomes a "Blakean archetype of an awesome natural force that is amoral as much as the people who are pitted against it in an eternal battle of survival," especially in the episode in which a tiger is being attacked by the villagers, but I would argue that it also serves as an antagonist of sorts, given Ghosh's environmental justice concerns about threatened human beings whose lives and livelihoods are endangered because of man-eating tigers or state-led tiger conservation policies (136). This ambivalent figuration of the tiger undermines the ecocentric perspective of the novel, to an extent, and illustrates how even though environmental justice and ecocentric perspectives may coexist within the same text, it is hard to balance or reconcile them in the ways that Buell hopes for.

Ghosh returns to his ecocentric concerns in the climactic closure of his novel. The narrative ends with a huge cyclone in which Fokir loses his life. What stands out in Ghosh's minute description of the cyclone and the devastation it wreaks in the region is that humans, animals, and the forest are equally susceptible to the dangers of high waters. The island of Garjontola has been almost entirely submerged by the tidal wave engendered by the storm while all living creatures fall prey to the fury of the storm. Piya describes a flock of exhausted white birds and a tiger in similar language:

> *One of the birds was so close she was able to pick it up in her hands: it was trembling and she could feel the fluttering of its heart. Evidently the birds had been trying to stay within the storm's eye.* . . . She saw a tiger pulling itself out

of the water and into a tree on the far side of the island. *It seemed to have been following the storm's eye, like the birds, resting whenever it could. . . .* (321, my emphasis)

Ghosh underscores the analogous response of the birds and the tiger: both try to stay within the eye of the storm, and both are extremely frightened of the high waters set loose by the cyclone, much like the two vulnerable human figures, Fokir and Piya, who find themselves in the midst of the storm. By bringing home the leveling effect of the storm as humans and nonhumans, alike, become victims of the cyclone's fury, the closure of the narrative, once again, accentuates Ghosh's ecological commitments. The novel views humans, tigers, dolphins, and birds, among other life forms, as an intrinsic part of the wider ecosphere of the Sundarbans and illustrates the interrelatedness and the shared fate of humans and other living beings, *including tigers.*

Moreover, the very form of the novel serves to reinforce Ghosh's ecocentric preoccupations. As Jens Martin Gurr notes, the novel is "profoundly imprinted with the characteristics of the landscape it unfolds from," which is a "wetland" comprised of both land and water (79, 60). (The Sundarbans is a unique and distinctive biotic space in which the land emerges only at low tide and thousands of acres of the forest disappear every day.) The dichotomy of land and water, ebb and flood, motivates the entire narrative of the novel.[29] As we have seen, the narrative alternates between two subplots: the social history of this land with the focalization on Kanai or mediated through Kanai, and Piya's quest for the Irrawaddy dolphin in the waters of the tide country. Piya is torn between Kanai and Fokir with Kanai representing the land—more specifically, urban metropolitan places—and Fokir the water. Indeed, the very structure of the novel relies on the particularities of the natural environment of the Sundarbans with the book being divided into two parts: "Ebb" and "Flood." In keeping with the notion of the ebb tide that gives life to the land, the penultimate chapter of the first section, titled "Dreams," has Nirmal recounting how the refugees have managed, despite remarkable odds, to make Morichjhapi habitable in a matter of weeks by pushing the water back. This is followed by a chapter titled "Pursued" in which Piya and Fokir are attacked by a crocodile in the water; in many ways this chapter anticipates the dangers of the second half of the novel, which is appropriately titled "The Flood: Jowar." This latter part of the novel details the defeat of the refugees despite all their efforts to resist eviction—the "flood" here functions as a symbolic flood—and it ends with the literal flood, as we have seen, which culminates in the death of Fokir and leads to much destruction in the tide country. Thus, Ghosh uses the distinct natural environment of the tide country to give shape to his novel and to articulate his ecocentric concerns.

TROUBLING CLOSURES

Interestingly, despite the devastating storm in which Fokir loses his life and
Kanai loses Nirmal's notebook, the novel ends on a somewhat upbeat note
with the two central characters—Kanai and Piya—discovering a new found
sense of responsibility and place-attachment to the tide country and its inhabitants, very much in the mode of what has been termed "vernacular cosmopolitanism."[30] Kanai, the privileged insular city dweller, is able to see himself through Fokir's eyes—in a rare moment of rendering the self-uncanny—
as the outsider, the representative of all those people who had killed his
mother. It was as though "he were seeing not himself . . . but a great host of
people. . ." (270).[31] This heightened moment of self-realization is followed
by a terrifying surreal encounter with a tiger in an otherwise strictly realist
text. The novel never clarifies whether the tiger was real or conjured up by
Kanai; what's more important is the marked change in him that is brought
about by the confrontation with Fokir and the tiger.

Eventually, Kanai decides that he will act as a "secondary witness" to the
story of Morichjhapi because Nirmal's notebook could not make its way to
the world.[32] It gets lost in the storm and becomes a metaphor for how the
Morichjhapi episode remains lost to the annals of history. Indeed, Nirmal is
well aware that he does not have the ear of an unheeding world given his
marginal existence in the tide country, which is why he addresses his testimony to Kanai: "I feel certain you will have a greater claim to the world's ear
than I ever had" (230). In keeping with his recently discovered commitment
to the history of the tide country and to its people, Kanai decides to move to
Kolkata to write the story of Nirmal's notebook, which is in effect the narrative of this novel.

Similarly, the novel traces the powerful impact of the tide country on
Piya. Initially, she is only interested in her research project—she views the
dolphins as an object of her scientific quest—but her involvement with Fokir
and her interactions with Kanai lead her to take on an ethical responsibility
toward the subaltern inhabitants of the tide country. She decides that she will
live in Lusibari for some time and continue with her conservation work, but
this time with a difference. She wants her work to be conducted under the
sponsorship of Nilima's trust so that she can work with the local fishermen.
A chastened Piya tells Nilima, "I don't want to do the kind of work that
places the burdens of conservation on those who can least afford it" (327). As
Rajender Kaur observes, Piya "matures from being a blinkered conservation
biologist focused only on studying the Oracella [sic] to a more progressive
environmentalist" (132). Rather than viewing the dolphins in isolation as a
particular subspecies that must be rescued, she comes to see the dolphins as
an intrinsic part of the larger ecosphere of the Sundarbans along with the
indigent human denizens. Ghosh himself notes how "a single-species ap-

proach to preservation is increasingly under question the world over" ("Wild Fictions" 73).

The closure of both Kanai and Piya's narrative trajectories seems to suggest in somewhat disturbing ways that "the emancipatory possibilities of an interventionist ethics reside within the very structures of middle-class privilege" (Tomsky 64), while the figure of the subaltern (both Fokir and Kusum) can only be sacrificed to summon Ghosh's (elite) readers out of their passivity and stupor in order to act ethically in the world.[33] Accordingly, Kusum's story ends in her death and probable rape, whereas Fokir sacrifices himself to save Piya from the storm. Although both endings are entirely plausible, they foreclose any possibility of subaltern agency within the terms of Ghosh's narrative. (Perhaps Fokir must die because Ghosh does not know how to resolve the triangulated romance at the center of his novel.) The fact that the subaltern figures, in effect, are written out by the text's closure is extremely troubling for a novel that is committed to creating the Sundarbans as a "place" for its cosmopolitan readership and to sustaining the diversity of its various life forms.

Despite its problematic closure, the novel remains, to my mind, a powerful work of environmentalism that brings together environmental justice questions with ecocentric commitments in compelling and persuasive ways. It allows us to see that an environmentalism of place does not necessarily need to be at odds with an environmentalism of slow violence and displacement.

NOTES

1. To be sure, Nixon's book is not the first to make an intervention within the field of "postcolonial ecocriticism." The dialogue between the fields of postcolonialism and environmental literary studies had been occurring for some time before Nixon's 2011 publication. The 2005 collection *Caribbean Literature and the Environment*, edited by DeLoughrey, Gosson, and Handley, is the first major volume in this field.

2. Bioregionalism entails "responsiveness to one's local part of the earth whose boundaries are determined by a location's natural characteristics rather than arbitrary administrative boundaries" (Parini, cited in Nixon 238).

3. I draw upon Nixon's insightful and layered use of the term "apprehension" here as opposed to the more common phrase "making visible" in the context of literary testimonies. As Nixon writes, the term *apprehension* merges together the domains of perception, emotion, and action. To apprehend—arrest, mitigate—slow violence requires making it apprehensible to the senses through immediate sensory perception, but also through the work of scientific and imaginative testimonies, which can help to make the "unapparent appear" by "humanizing drawn-out threats inaccessible to the immediate senses" (14–15).

4. Humanist geographers suggest that once a space is endowed with value and meaning it becomes a place. Both Nixon and Buell recognize that there is nothing intrinsically good or bad about place attachment, but where the former highlights the dangers of commitment to a particular place, the latter is more invested in thinking about the potentials of place as a "resource" for generating environmental concern (56).

5. For an account of how the spectrum of what counts as environmentalism has opened up, see Nixon (4–5).

6. As Annu Jalais points out, the term *bhadralok* (gentle folk) is widely used in Bengal and refers to the "rentier class who enjoyed tenurial rights to rents from land appropriated by the Permanent Settlement." This was a class that did not work the land but lived off the rental income generated from the land. They shunned manual labor and saw this as key to the difference between them and those they considered as social inferiors. The term carries with it connotations of Hindu, upper-caste privilege, and of landed wealth. See "The Sundarbans: Whose World Heritage Site?" (7).

7. Although I follow scholarly convention in using the upper case for the 1947 partition, I take seriously van Schendel's injunction that we must not essentialize the Indian partition as a unique and peerless event; rather we must take a comparative approach even within South Asia for the partition of 1947 stands between the creation of the separate colonial state of Burma in 1937 and the division of Pakistan in 1971. See van Schendel (26–28).

8. I have drawn on Gyanesh Kudaisya's "Divided Landscapes" for this account of the Dandakaranya camp and the Morichjhapi massacre. See also Annu Jalais's "Dwelling in Morichjhapi."

9. Taking Edward Said's call to think beyond the canonical Western literature of exile as its point of departure, this chapter is part of a larger work that examines the mass displacements that emerged in the wake of the partitions of India and Pakistan in 1947 and 1971—and the identities and conflicts that they spawned—in terms of their enduring legacies. In an effort to provide a more capacious conceptualization of South Asian diasporas, I aim to extend the meaning of "exile" and "diaspora" by moving past metropolitan spaces and diasporas in the West to focus on migrant communities engendered by the partitioning of nation-states in the subcontinent—and their descendants—that have often fashioned themselves as diasporas or nowhere people even as the two new states positioned these migrants as "proxy citizens" who were coming home to their respective nations. I borrow the notion of homecoming and "citizens by proxy" from Willem van Schendel's insightful book, *The Bengal Borderlands*. See also Said's "Reflections on Exile."

10. Jacques Derrida, *Demeure: Fiction and Testimony*. For more on how Derrida demonstrates the entanglement of fiction and testimony, see my *Limiting Secularism*, chapter 5.

11. However, to dismiss Nirmal's actions as emanating merely from his feelings for Kusum would be to trivialize them. As Terri Tomsky writes in her excellent essay on affect and anxious witnessing in *The Hungry Tide*, Nirmal's behavior is an "indication of his complex affective experiences that arise from her [Kusum's] proximity and his exposure to the galvanizing atmosphere and energies of Morichjhapi's community" (59). Hence, both factors are equally important in energizing Nirmal. Moreover, his passions for Kusum remain unrequited and are further "characterized by an ascetical degree of self-sacrifice" (60).

12. For more on these different representational convention, see Said's *Reflections on Exile*.

13. It is important to keep in mind that Bengali Dalit works like Sudhir Ranjan Haldar's *Aranyer Andhokarey* (From the Darkness of the Jungles) and Nakul Maliick's *Khama Nei* did address the plight of the Dandakaranya refugees before *The Hungry Tide*. That these works are hardly known to a Bengali reading public—let alone a wider South Asian English-reading public—is indicative of how certain literary testimonies see the light of day whereas others remain invisible and lost despite their efforts to bear witness. This awareness is also indicated within the novel when Nirmal addresses his testimony to Kanai to make sure that it is transmitted to other readers. I am grateful to Brati Biswas for introducing me to these Bengali Dalit narratives.

14. See Paul Greenough's "Pathogens, Pugmarks, and Political 'Emergency'" for a very insightful and informative account of Project Tiger.

15. However, it is important to keep in mind that anthropocentrism encapsulates a range of positions from the strong belief that human interests should be elevated over and above those of all other species to the more pragmatic acceptance that ecocentrism and biocentrism must be constrained by anthropocentric considerations. See Buell's *Future* (134).

16. Indeed, the "Principles of Environmental Justice," put forward by participants in the 1991 First National People of Color Environmental Leadership Summit held in Washington, D.C., overlap quite a bit with traditional environmentalism including belief in the sanctity of "Mother Earth" and the interdependence of all species (cited in Buell, *Future* 114). Ghosh's

environmental justice concerns are also very evident in his essay "Wild Fictions" in which he provides an incisive critique of the romantic idea of a pristine nature uncontaminated by human beings. Like Guha, he interrogates the idea of national parks and wildlife sanctuaries that have led to the creation of "environmental refugees, who have been evicted in the process of creating the parks" (62). He suggests that the inhabitants of these settlements—who are among some of the poorest people in India—have paid the costs of protecting nature, whereas certain sections of the urban middle class have reaped the rewards (64).

17. In *Writing for an Endangered World*, Buell points to an "ethical/aesthetic schism" that has come about in literary narratives as a consequence of the anthropocentric-ecocentric binary such that "the imagined miseries of beasts and humans generate their own specialized genres" and commentators (220). He reads Mahasweta Devi's Bengali novella *Pterodactyl, Puran Sahay, and Pirtha* and Barbara Gowdy's novel *The White Bone* as symptomatic of this anthropocentric-ecocentric split in literature. If Pterodactyl is "one of the most trenchant and challenging fictions of environmental justice ever written" in its effort to highlight the plight of tribal people or Adivasis in India, then *The White Bone* is a paradigmatic biocentric text in its bold attempt to imagine how elephants think and feel. However, Buell claims that each literary narrative marginalizes the other type of misery. In Gowdy's novel, humans are portrayed as grotesque and demonic killers. While Mahasweta Devi's novella is not as divisive because it suggests that the adivasis' condition is linked to the devastation of the non-human environment, nevertheless, it marginalizes the suffering of the non-human other. For the most part, then, Buell suggests "the two texts adhere to specialized forms of ethical-aesthetic extensionism that exclude each other's concerns" (234). In conclusion, he asks: Can one hope to find in modern literature works that do not fall into this binary of either an anthropocentric or an ecocentric perspective?

18. Postcolonial ecocriticism is typically recognized as part what has come to be known as third-wave ecocriticism, as Scott Slovic explains in "The Third Wave of Ecocriticism: North American Reflections on the Current Phase of the Discipline" (*Ecozon@* issue 1.1, April 2010). In many ways, I see Ghosh's novel as part of a postcolonial ecocritical impulse in its efforts to highlight the plight of dispossessed humans in the Global South and its (simultaneous) concern about the natural world of the Sundarbans. For more on postcolonial ecocriticicsm, see such books as the DeLougrey, Gosson, and Handley volume mentioned above and Graham Huggan and Helen Tiffin's *Postcolonial Ecocriticism: Literature, Animals, Environment* (2010), Bonnie Roos and Alex Hunt's *Postcolonial Green: Environmental Politics and World Narratives* (2010), Pablo Mukherjee's *Postcolonial Environments: Nature, Culture, and the Contemporary Indian Novel* (2010), and Elizabeth DeLoughrey and George B. Handley's *Postcolonial Ecologies: Literatures of the Environment* (2011).

19. In his more recent book, *The Future of Environmental Criticism*, Buell identifies two Australian works, "Celebrators 88" a poem by aboriginal author Kevin Gilbert, and "Inventing the Weather," a novella by Queensland writer Thea Astley, as texts that show how "environmental justice commitments can coexist with an ecocentric persuasion" (124). My reading of *The Hungry Tide* makes a similar argument. Ironically, in their review of ecocriticism, Buell, Heise, and Thornber view *The Hungry Tide* only in terms of its critique of wildlife preservation and don't attend to the text's ecocentric concerns.

20. On Morichjhapi and East Bengali refugees, see the essays by Nishi Pulugurtha, Rituparna Roy, and Supriya Choudhury. See also Amrita Ghosh's forthcoming work on the novel. On works that attend to the novel's ecocritical—especially ecocentric—concerns, see the works by Rajender Kaur, Serenella Iovino, Arnapurna Rath and Milind Malshe, and Jens Martin Gurr for an illustrative sample. Martin Gurr's essay is instructive for its attention to questions of form and how the form of the novel shores up Ghosh's ecocritical preoccupations. However, Martin Gurr tends to equate nature writing and environmental literature in keeping with first wave ecocriticism; for example, invoking the Morichjhapi incident briefly, he says the novel has much to say on issues of environmental justice and human rights, but that is "another matter" dealt with by the other essays in this collection. His focus is on "environmental concerns"—the implication being that environmental literature does not include within its realm narratives of environmental justice. This is in sharp contrast to Rob Nixon's selection of texts in *Slow Violence*. Rajender Kaur's essay, "Home is Where the Oracella [sic] Are: Toward a New

Paradigm of Transcultural Ecocritical Engagement in Amitav Ghosh's The Hungry Tide," is one of the first and more nuanced ecocritical readings of the novel. However, Kaur tends to focus almost entirely on Piya's narrative trajectory to the extent of privileging her centrality to the novel's broader environmental concerns. Hardly any attention is paid to the importance of the Bon Bibi story in the novel in terms of how it allows the islanders to forge a relationship with the natural environment and to respect the space of nonhuman others. And while Kaur insightfully suggests that "a life of dignity, for even the most marginalized citizens of the world, alongside socially responsible environmental policies that further preserve the unique biodiversity of our planet, need not be mutually exclusive goals," she, too, tends to view environmental justice issues as outside the ambit of environmentalism or at odds with it when the very spectrum of what we consider as environmentalism and environmental literature has expanded. My approach sees both environmental justice and ecocentric or biocentric texts as central to what we constitute as "environmental writing."

21.   Indeed, for much of its colonial history, the Sundarbans was designated as an unpopulated "wasteland" and the modern inhabitation of these islands only goes back to the early twentieth century when Sir Daniel Hamilton bought ten thousand acres of the tide country from the colonial government to implement his dream of a casteless, classless society to be run by cooperatives by inviting people to come and settle in these islands (Ghosh, 43–44). At the same time, Ghosh also underlines how the very features of the natural environment that make the islands hostile to human settlement also made them a refuge and a sanctuary for landless and homeless people over the decades. (I am indebted to my student Margaret Hass for this insightful point.)

22.   For an incisive critique of this strong anthropocentric philosophical tradition, see Jacques Derrida's *The Animal that therefore I am.*

23.   In Annu Jalais's version of the legend, Bon Bibi is born in the Sundarbans, but she goes to Medina upon hearing Allah's call to free the "land of the eighteen tides" from Dokhin Rai to receive Fatima's blessings, and eventually returns to the Sundarbans (*Forest of Tigers*, 70-75).

24.   Jalais points out that Dokhin Rai is supposed to be a Brahmin sage who takes on the guise of a tiger in order to quench his desire for human flesh (70). The legend thus amalgamates the figure of the tiger and the Brahmin in very interesting ways, especially given that most of the worshippers of Bon Bibi are forest workers who belong to the lower castes.

25.   Jalais explains that the Bangla word *jongol* differs from the English word "jungle" and its connotations of a dense thicket of tropical vegetation. *Jongol* connotes wilderness, specifically the sphere of nonhumans in the Sundarbans (233).

26.   I am grateful to my student Satyendra Singh for picking up on the resonance of the shared phrase.

27.   The question of who is indigenous to the Sundarbans, given its recent history of settlement, of course is open to debate. Jalais observes that the islanders think of themselves as migrants, but they are not invested in some imaginary homeland left behind in their ancestral places of origin. They see themselves "as migrants *to* the Sundarbans, rather than migrants *from* somewhere else." Thus, they can all lay equal claim to the Sundarbans (*Forest of Tigers* 182).

28.   In recent years, there has been much talk of promoting and developing the Sundarbans islands as a tourist destination for urban nature and wildlife lovers given the unique ecosystem of the Sundarbans and its claim to being the only mangrove forest in the world to be inhabited by tigers. In 1987, the Sundarbans was declared a World Heritage Site by the International Union for the Conservation of Nature. However, Jalais argues that these recent wildlife conservation discourses tend to omit humans from their discursive constructions of the islands, thereby increasingly alienating the islanders from their environment and from tigers in particular. She asks: "how can we talk of or represent the Sundarbans without taking into account its people and their understandings of place?" Unless, we learn to provide people with a stake in conservation projects or in other kinds of environmental efforts, we will only alienate these communities who feel that their lives are of less value than tigers, and state-initiated conservation policies will be doomed to failure. This sentiment of dispossession is especially evident in the ways in which the Morichjhapi incident is remembered by the Sundarbans islanders many years after the event. In a fascinating instance of anthropomorphization, the islanders reiterate how tigers became more arrogant after Morichjhapi because they felt that the state was on their

side, and how they, the islanders, are only seen as "tiger-food" by the state and by elite middle-class Bengalis. See "Dwelling in Morichjhapi."

29. This section draws upon, modifies, and elaborates on some of Martin Gurr's formulations on the importance of form in the novel.

30. See Homi Bhabha's conceptualization of vernacular and minority cosmopolitanism. I am grateful to Tarun Saint for reminding me of the parallels of this strand of the novel with vernacular cosmopolitanism.

31. On the notion of rendering the self-uncanny, see Gayatri Spivak's *Death of a Discipline* and my *Limiting Secularism*, chapter 2.

32. I borrow the notion of a secondary witness from Dominic LaCapra. See chapter 3 of *Limiting Secularism* for an elaboration of this concept.

33. I agree with Tomsky's reading of the conclusion, but I depart from the positive valence she puts on it.

WORKS CITED

Bauman, Zygmunt. *Wasted Lives: Modernity and its Outcasts*. Malden, MA: Polity, 2004. Print.

Bhabha, Homi. "Unsatisfied: Notes on Vernacular Cosmopolitanism." *Text and Nation: Cross-Disciplinary Essays on Cultural and National Identities*. Ed. Laura Garcia-Moreno and Peter C. Pfeiffer. Columbia, SC: Camden House, 1996. 191–207. Print.

Buell, Lawrence. *Writing for an Endangered World: Literature, Culture, and Environment in the U.S. and Beyond*. Cambridge, MA: Belknap Press of Harvard UP, 2001. Print.

———. *The Future of Environmental Criticism: Environmental Crisis and Literary Imagination*. Malden, MA: Blackwell Publishing 2005. Print.

Buell, Lawrence, Ursula K. Heise, and Karen Thornber. "Literature and Environment." *Annual Review of Environment and Resources* 36 (2011): 417–40. Print.

Chaudhury, Supriya. "A Sense of Place: Book Review of *The Hungry Tide*." *Biblio* July-August 2004. Web.

DeLoughrey, Elizabeth, Renee K. Gosson, and George B. Handley, eds. *Caribbean Literature and the Environment: Between Nature and Culture*. Charlottesville: U of Virginia P, 2005. Print.

Derrida, Jacques. *Demeure: Fiction and Testimony*. Trans. Elizabeth Rottenberg. Stanford: Stanford UP, 2000. Print.

———. *The Animal That Therefore I am*. Trans. David Wills. New York: Fordham UP, 2008. Print.

Giri, Bed and Priya Kumar. "On South Asian Diasporas." *The South Asian Review* 32.3 (2011): 11–26. Print.

Ghosh, Amitav. *The Hungry Tide*. London: HarperCollins, 2004. Print.

———. "Confessions of a Xenophile and Wild Fictions." *Outlook* (December 2008): 32–88. Print.

———. "Wild Fictions." *Outlook India.com*. 22 December 2008. http://www.outlookindia.com/article.aspx?239276. Accessed November 30, 2014.

Greenough, Paul. "Pathogens, Pugmarks, and Political 'Emergency.'" *Nature and the Global South*. Ed. Paul Greenough and Anna Lowenhaupt Tsing. Durham, NC: Duke UP, 2003. 201–30. Print.

Guha, Ramachandra. "Radical American Environmentalism and Wilderness Preservation: A Third World Critique." *Varieties of Environmentalism: Essays North and South*. Ed. Ramachandra Guha and J.Martinez-Alier. New Delhi: Oxford UP, 1998. 92–108. Print.

Gurr, Jens Martin. "Emplotting an Ecosystem: Amitav Ghosh's *The Hungry Tide* and the Question of Form in Ecocriticism." *Local Natures, Global Responsibilities: Ecocritical Perspectives on New English Literature*. Ed. Laurenz Volkmann, Nancy Grimm, Ines Detmers, and Katrina Thomson. Amsterdam and New York: Rodopi, 2010. 69–80. Print.

Iovino, Serenella. "Ecocriticism and a Non-Anthropocentric Humanism." *Local Natures, Global Responsibilities: Ecocritical Perspectives on New English Literature*. Ed. Laurenz Volkmann, Nancy Grim, Ines Detmers, and Katrina Thomson. Amsterdam and New York: Rodopi, 2010. 29–53. Print.

Jalais, Annu. "Dwelling in Morichjhanpi: When Tigers Became 'Citizens', Refugees 'Tiger-Food'." *Economic and Political Weekly* (April 23, 2005): 1757–62. Print.

———. "The Sundarbans: Whose World Heritage Site?" *Conservation and Society* 5.3 (2007): 1–8. Print.

———. Jalais, Annu. *Forest Of Tigers: People, Politics And Environment In The Sundarbans* . New Delhi: Routledge, 2009. Print.

Kaur, Rajender. "Home is Where the Oracella Are: Toward a New Paradigm of Transcultural Ecocritical Engagement in Amitav Ghosh's *The Hungry Tide.*" *ISLE: Interdisciplinary Studies in Literature and Environment* 14.1 (Summer 2007): 125–41. Print.

Kumar, Priya. *Limiting Secularism: The Ethics of Coexistence in Indian Literature and Film.* Minneapolis: U of Minnesota P, 2008. Print.

Kudaisya, Gyanesh. "Divided Landscapes, Fragmented Identities: East Bengal Refugees and their Rehabilitation in India." *Singapore Journal of Tropical Geography* 17.1 (1996): 24–39. Print.

LaCapra, Dominic. "Trauma, Absence, Loss." *Critical Inquiry* 25 (1999): 696–727. Print.

Mishra, Vijay. "The Diasporic Imaginary: Theorizing the Indian Diaspora." *Textual Practice* 10.3 (1996): 421–47. Print.

Nixon, Rob. *Slow Violence and Environmentalism of the Poor.* Cambridge, MA: Harvard UP, 2011. Print.

Pulugurtha, Nishi. "Refugees, Settlers, and Amitav Ghosh's *The Hungry Tide.*" *Local Natures, Global Responsibilities: Ecocritical Perspectives on new English Literatures.* Ed. Laurenz Volkmann, Nancy Grimm, Ines Detmers, and Katrina Thomson. Amsterdam and New York: Rodopi, 2010. 81–89. Print.

Roy, Rituparna."*The Hungry Tide*: Bengali Hindu Refugees in the Subcontinent." *The Newsletter* 51 (2009). Web.

Spivak, Gayatri Chakravorty. *Death of a Discipline.* New York: Columbia UP, 2003. Print.

Said, Edward. *Reflections on Exile and Other Essays.* Cambridge, MA: Harvard UP, 2002. Print.

Tomsky, Terri. "Amitav Ghosh's Anxious Witnessing and the Ethics of Action in *The Hungry Tide.*" *Journal of Commonwealth Literature* 44 (2009): 53–65. Print.

Van Schendel, Willem. *The Bengal Borderland: Beyond State and Nation in South Asia.* London: Anthem, 2005. Print.

Volkmann, Laurenz, Nancy Grimm, Ines Detmers, and Katrina Thomson, eds. *Local Natures, Global Responsibilities: Ecocritical Perspectives on the New English Literatures.* Amsterdam and New York: Rodopi, 2010. 69–80. Print.

I am grateful to Barbara Eckstein and Mary Lou Emery for their thoughtful comments on an earlier version of this chapter. Many thanks to Scott Slovic and Vidya Sarveswaran for their very helpful editorial suggestions.

Chapter Two

"The Land Was Wounded": War Ecologies, Commodity Frontiers, and Sri Lankan Literature

Sharae Deckard

Anglophone Sri Lankan literature is saturated by spatialized registrations of ecology. Gothic eco-topoi such as the spectral *waluwe* (plantation house), the ecophobic, fecund jungle, and the toxic gothic of militarized waste-scapes, recur throughout the fiction of writers such as Punyakante Wijenaike, Jean Arasanayagam, Romesh Gunesekera, Ameena Hussein, and Roma Tearne, mediating the history of the socio-ecological production of nature through plantation monocultures, paradise tourism, and military territorialization. These eco-tropes figure ecologies, which have been subjected to multiple reterritorializations, so that literary representations of landscapes become palimpsests of multiple socio-ecological histories, drenched in accumulated violence. Thus, to name one example, Jean Arasanayagam, a Burgher poet, playwright, and novelist whose oeuvre offers one of the most sustained and powerful engagements with the twenty-six-year Sri Lankan civil war, repeatedly depicts the heavily contested environments of the North, oscillating violently between military control by The Liberation Tigers of Tamil Eelam (LTTE) and Sri Lankan state security forces, as uncannily nourished by the blood of "ethnic cleansing," "massacres and assassinations," "disappearances, torture, death" (*All* 5). The use of tropes of blood-soaked soil throughout her work to describe a "wounded land" implies an ecophobic dimension—the horror of human violence projected onto the extra-human landscape, with a certain added resonance from the colonialist imagination of the "deadly" fecundity of the tropics—but transcends a merely instrumentalist use of environmental imagery, in that the deathscapes she imagines correspond to "war ecologies" produced by heavy militarization, deforestation,

35

shelling, use of pesticides and gas, destruction of irrigation infrastructure, and the burning of crops, villages, and fields to prevent cultivation and to strip enemy forces of shelter and sustenance.

War ecology, as I use it here, refers to the radical reorganization of nature in order to eliminate the possibility of life—whether that of the guerrilla, the villager, or the soldier—through the annihilation of the interdependent relations with extra-human nature that sustain human life. It is an inverted ecology which cultivates death, rendering nature into militarized zone. In her poem "Goyaesque Etchings from 'The Disasters of War'," Arasanayagam describes villages emptied of agriculture, their inhabitants slaughtered, impressed into military service, or fled to refugee camps, their fields burnt. Indigenous peasant agriculture, the cultivation of rice from *chena* plots slashed out of the jungle, is replaced by the cultivation of "new harvests" of "blood milk," as the villages return "to forest" (*Fault* 20). The peasants "want to go back to the golden earth . . . to sow our seed" (*Fault* 19), but find that their sacks which once carried paddy are now full of "the new blood harvest" (*Fault* 19). In her play "The Fire Sermon," the trope of a harvest of death is repeated as the characters Alice and Yama imagine that "all this blood must seep into the earth" (*Fault* 54) to produce ecology of death: "Everything is polluted now by death—water, air, plants. Malignant. Evil. Black mushrooms streaked with crimson spring up everywhere. Poisonous. Not like the pure white mushrooms that people gathered early in the morning. They are digging graves everywhere . . ." (*Fault* 55). Similarly, in the short story "The Journey," she describes Europe through the eyes of a Sri Lankan illegal immigrant fleeing the civil war, as a "reclaimed territory [that] rests on a foundation of skeletal remains: bones that branch out like a subterranean forest, the flesh nourishing the soil" (*All* 5). In linking the European Balkans to the contested territories of the Sri Lankan civil war, she deprovincializes the trope of the bloodied forest, subverting the exceptionalist myth of the "tragic island" by suggesting that no environments are ever ahistorical, untouched, or idyllic, whether in the Global South or northern capitalist countries. Instead, territories must be read in light of the human production of nature and conflict for resources through which they are inscribed and reinscribed.

Vinay Dharwadker has observed that the over determination of "race" and "land" in Sri Lankan and LTTE discourses creates a difficulty for both critics and writers: "Sri Lankan poets mostly confront a natural environment that is always already too thickly covered with the discourse, mythology, and architecture of the island's long-standing, disparate religions" (Dharwadker 277). Somewhat paradoxically, the dominant strand of Anglophone Sri Lankan criticism has been oriented toward producing critiques of the discourses of nationalist chauvinism, ethnic purity, myths of origin, and religious and political sectarianism which have fuelled the civil war. In doing so it has championed the use of "migrant," "hybrid," and "diasporic" postcolonial aesthetics to challenge such discourses, so that the environment has often failed to register as more than a discursive terrain or contested site of land-as-nation, despite the intensely ecological content of much of Sri Lankan writing. The place of environment in Sri Lankan aesthetics has frequently been read as political metaphor and imagery, but rarely as registering material socio-ecological relations, such as those of the plantation.

This tendency seems curious when compared to other literatures from island-states in the Global South whose environments and cultures have been formed by and continue to be deeply shaped by plantation monoculture and export dependency. The "environmental turn" in literary criticism has been particularly strong in studies of the Caribbean, whose archipelago has been indelibly marked by the boom-bust monocultures of sugar, tobacco, and other commodities, produced by brutal reliance on forms of enslaved and unfree labor, and wreaking violent ecological transformations in the radical simplification of nature required to produce extraction of commodities for large-scale export. Comparativism of literatures of plantation and commodity monocultures throughout the Global South—whether from the Caribbean, South Asia, Africa, or East Asia—offers huge potential for world-ecological criticism to uncover both structural analogies and aesthetic specificities which speak to the particularity of ecologies and cultures in different regions, while at the same time revealing the ways in which they are linked by their teleconnection to the capitalist world-ecology and world-market.

In colonial Ceylon, large-scale plantation occurred later than in the Caribbean, in the nineteenth century, but the last one hundred and seventy years of Sri Lankan history have been no less determined by the asymmetries of the full-fledged plantation economy established in the 1840s, which simultaneously modernized and underdeveloped Ceylon, creating an enduring economic structure designed to serve the interests of a planter class while rendering the country dependent on the export of a handful of commodity cash-crops—tea, rubber, and coconut—for consumption in the cores of the capitalist world-system—first in the imperial metropole, but now throughout the Global North. If war ecologies are singularly organized around the elimination of the ecological relations which sustain human life-worlds, plantation ecologies are organized around the singular production of commodities, a radical simplification of nature which eliminates other forms of life outside commodity monocultures. The war ecologies and death-worlds produced by the civil war between the Tamil Tigers and the state security forces are particular to Sri Lankan history, but they cannot be understood purely as the social product of ethnic, racist, and chauvinist discourses, but rather as indelibly rooted in the preceding environmental relations of plantation: the brutal reorganization of precapitalist nature-society that dispossessed indigenous peasantries, imported indentured labor, converted vast tracts of highland to cash-crop monocultures, and radically destabilized ecologies, while using racist divide-and-rule policies to control populations, partition land, and discipline labor. I will argue in this chapter that plantation in Sri Lanka should be understood in environmental historian Jason Moore's terms as an "ecological regime" inaugurated in the nineteenth century and persisting after independence to the present day, with ongoing deleterious consequences for Sri Lankan economy, environment, and society (Moore, *End* 392).

WORLD-ECOLOGY AND THE GLOBAL SOUTH

For Moore, "nature" and "society" are not independent units which should be integrated in criticism; rather they must be conceptualized dialectically as an *oikeios*, as a web of life incorporating "bundles" or "relations" of both human and "extra-human" nature, taking in the plethora of ways human and biophysical natures are intertwined at every scale from the microbiome, to the body, to city, to the world market, and the ways in which humans are actively engaged in but also constrained within manifold patterns of environment-making, from agriculture to cultural production to financialization. In other words, nature is not to be conceived dualistically as merely "resource bin" or "rubbish dump," even if the dominant tendency of capitalist rhetoric has been to imagine nature as either mine or sink, a source of seemingly "free gifts" to be plundered for profit. Moore adapts world-systems approaches in order to narrate capitalism as environmental history, and to chart the different phases of the "capitalist world-ecology" structurally differentiated and riven between cores and peripheries. He argues that environmental history should be understood not as the history of capitalism working "on" nature in order to produce profits, but rather of capitalism emerging "through" the periodic reorganizations of nature-society relations, which he terms "ecological regimes" and "ecological revolutions." Ecological regimes are the "relatively durable patterns of class structure, technological innovation, and the development of productive forces . . . that have sustained and propelled successive phases of world accumulation" (Moore, *End* 392).

According to Moore, the history of ecological regimes in (post)colonies throughout the Global South is characterized in particular by the sectoral relocations and rise and fall of "commodity frontiers," a term which he adapts from "commodity chains" to focus not merely on the chain of economic production as it extends from economic peripheries to cores where commodities are consumed and surplus value is concentrated, but rather to draw out the means by which manufacturing chains emerge through specific socio-ecological relations at the extractive beginnings of chains, where colonial empires and now transnational corporations maintain relations of exploitation and domination. Because plunder exhausts the noncommodified relationships that allow capital accumulation to proceed, capitalism is always in search of new commodity frontiers for extraction and appropriation. In Moore's studies of plantation monocultures in the Caribbean, the rapid appropriation of commodity frontiers in cash-crops such as sugar, undermines the socio-ecological conditions of profitability typically within fifty to seventy-five years in any given region, encountering not only biophysical limits to extraction such as soil exhaustion, blight, or superweed effects, but also the scarcities which "emerge through the intertwining of resistances from laboring classes, landscape changes, and market flux—all specific bundles of relations between humans and the rest of nature, specific forms of *oikeios*" (Moore *Wall* 46). Commodity monocultures, and the societies they produce, are governed by boom-bust phases, related not only to the oscillation of volatile commodity prices in the world-market, but by the biophysical limits to production. When biophysical webs of life are exhausted and particular commodity regimes are no longer able to produce ever-greater ecological surpluses for capitalist cores, the conditions of profit accumulation falter and provoke "ecological revolutions," characterized by the relocation of the frontier to new geographies, the intensification of existing forms of extraction, and the production of new technologies and modes.

Michael Niblett has argued for a comparative study of commodity regimes and frontiers, particularly the monocultures arising from plantation, where literary texts register the domination of local societies by specific commodities (Niblett 18). The cyclical rise and fall of commodity regimes enables a comparative approach that stresses the periodicity and geographical comparability of the registration of commodity regimes in literatures across the Global South. Texts registering the world-historical movement of coffee, tea, sugar, rubber, or opium commodities can be compared within national traditions, across regions such as the Caribbean or Latin America, or between macro-regions and continents, such as South Asia, East Africa, or East Asia in the case of tea. Within literary texts' registrations of cyclical regimes, differences in the registration of the periodicity of these cycles could be more finely distinguished between the "ecological revolution"—the moment combining the exhaustion of the previous regime with the radical reorganization of socio-ecological unities—and the "ecological regime," when socio-ecological relations are reconsolidated. The world-historical moment could also be varied, tracing the progression of regimes in a particular national context, tracking the progressive displacements of crises across geographies as commodity frontiers are sectorally relocated, or comparing different epochal and developmental phases of capitalism.

Tea and rubber both demonstrate the hallmarks of regional boom-and-bust cycles and sectoral relocation which characterize commodity regimes. In Sri Lanka, socio-ecological relations are indelibly marked by the commodity frontiers corresponding to tea, rubber, and coconut, originating in colonial plantation but continuing through the eras of independence, nation-alization, and subsequent re-privatization, and by the subsequent reorganiza-tions of society-nature manifested during the civil war and its ongoing con-flict over territory, labor, and resources. Contemporary Sri Lankan literature is permeated with representations of the ecological regimes and revolutions corresponding to plantation commodity frontiers and war ecologies; register-ing the rise and fall of tea, rubber, and coconut commodity regimes; the desacralization, deforestation, and toxification of jungle and dry zone ecolo-gies through militarization; the slow violence of environmental refugeeism and stationary dispossession; and the complex restructuring of previous re-gimes and creation of new commodity frontiers such as mass prawn aquacul-ture and paradise tourism during the neoliberal period. In what remains of this chapter, I will offer some preliminary explorations of commodity aes-thetics, particularly those corresponding to tea, by way of sketching the huge terrain that remains for future world-ecological criticism to excavate.

SRI LANKA, PLANTATION, AND COMMODITY FRONTIERS

I will begin with a more in-depth examination of commodity frontiers and plantation in Sri Lanka. Although the Dutch had cultivated small-scale cinna-mon plantation and gem-pit mining along the coastal regions before the advent of British colonialism, modern intensive plantation did not commence in Ceylon until the 1840s, when the British finally conquered the interior highlands of the Kandyan kingdom and began an intensive deforestation of the hills, systematically burning off the forest mantle in order to convert them to mass coffee cultivation. Dawood calls this the "second phase" of the "agro-extractive strategy of plantation development," originated in the Carib-bean and now transported to South Asia, where the British hoped to replicate the lucrative cash-crop monocultures of the West Indies, whose profits were dwindling after the abolition of slavery (Dawood 17). Scottish "tropical pio-neers" had tried first to convert the highlands to a mass grazier economy similar to the Scottish highlands and Irish countryside, until rinderpest deci-mated the imported cattle, and planters switched to coffee instead, unleashing a "coffee land rush" compared by colonial administrator James Tennent to the gold rushes of Australia and California, "with the difference that the enthusiasts in Ceylon, instead of thronging to disinter, were hurrying to bury their gold" (qtd. in Dawood 31).

Exploiting the feudal custom of *rajakariya*, colonial administrators used unpaid, forced labor by Sinhalese to clear the jungle and build over three thousand miles of macadamized roads in the plantation districts, "arteries" connecting the central hills to the ports of Colombo, Galle, and Trincomalee, which acted like "colossal drains carrying a people's wealth and birthright to foreign lands" (Dawood 25). The Crown Land Encroachment Ordinance of 1840 precipitated a massive expropriation of the peasantry, seizing over 90 percent of cultivable land, deeming it "waste-land" through a process of "legislated robbery" (Dawood 4) and selling it for a profit to speculators and planters, converting over ninety thousand acres into coffee plantations, in what James Webb has called the most extensive conversion of rainforest into tropical plantation agriculture anywhere in the British Empire (Webb 2). The rapid integration of Ceylon into the world-economy shattered the socio-ecological unities of the indigenous feudal culture: "King Coffee ruled and people rotted" (Dawood 19). The dispossessed peasantries were subjected to harsh grain and paddy taxes, in the attempt to drive them off their small plots into wage-labor on the plantations. Their water tanks were sold to planters, and soil erosion caused by mass denudation of the rainforest silted the rivers and remaining paddy lands, while their livestock were shot for trespassing on plantation land. However, after the traumas of *rajakariya* forest-clearing and dispossession, the Sinhalese displayed a stubborn resistance to plantation labor, even when exposed to starvation, preferring *chena* farming and bare subsistence to coerced labor. Demonizing the Sinhalese as "lazy" peasants mysteriously transformed from industrious agriculturalists, and fearing political unrest if they forced the issue, the planters imported indentured migrant Tamil labor from south India, repatriating the "surplus" population suffering from the Madras famine due to the combination of harsh colonial policies, taxes, and drought.

At the height of the coffee boom, production overtook West Indian coffee production until the sudden bust in 1869, when a coffee fungus known as "dreadful Emily" (*Hemileia vastatrix*) spread across the entirety of the monoculture, which had been rendered vulnerable to blight and pests due to the loss of biodiversity, soil exhaustion, destruction of shade trees, and water-table contamination. As quickly as it had emerged, the coffee frontier collapsed, and the planters turned to the new frontier of tea, following the successful experiments of young planter James Taylor and the agricultural research of the Royal Botanical Gardens at Peradeniya. The cool, hilly highlands were converted for the introduction of the tea shrub (*Camellia sinensis*), an evergreen which could be grown year-round at a variety of altitudes, inaugurating a new round of landslides, soil erosion, and elimination of wildlife and biodiversity after the large-scale clearing of the forest ecologies. In the southwest, the moist, tropical lowlands were sold to British entrepreneurs for the development of rubber (*Hevea brasiliensis*, smuggled from colonial Brazil by James Wickham) and the remaining lands were monopolized for coconut palm by the local Ceylonese entrepreneurs paying "arrack rents" in exchange for domestic profits on arrack liquor, which they sold to workers newly inducted into a wage-economy. Indigenous entrepreneurs were mostly prohibited from or unable to compete with the colonial tea and rubber estates,

but the small *Karava* elite—the so-called coconut kings—made their fortunes in the arrack trade, and were able to diversify their plantations after
independence, when they became the nation's political elites. As Patrick
Peebles wryly observes, "the wealth of the Karava capitalists was indeed
distilled from the nectar of the coconut palm" (Peebles 81).

Within tropical commodity chains such as sugar, cocoa, coffee, and tea
which originated as colonial crops, commodities are usually shipped out of
peripheral producing regions as soon as the products become storable; the
majority of processing, marketing, and "value-creation" takes place in the
cores (Talbot 707). However, unlike coffee and cacao which can be inter-
cropped by small-holders in more ecologically sustainable conditions and
processed in smaller batches with rudimentary technology, black fermented
tea is labor and capital-intensive and relies on economies of scale for profits,
which means that it has remained a plantation crop, grown in mass, rationalized estates, controlled first by colonial planters, then by transnational corporations. Like the sugar mills of the West Indies that require large volumes of
cane to crush in order to operate "efficiently," tea factory-farms require a
large continuous supply of freshly plucked tea (Talbot 713). It has to be
harvested year-round, so requires a stable labor force, rather than seasonal
migrant labor. The delicate tips of the new shoots are picked daily by hand,
while initial processing—withering, crushing, fermenting, and drying—has
to take place immediately at the site of harvesting, before "made tea" can be
exported overseas. In Sri Lanka, the division of labor for tea estate workers
has been strongly gendered since colonial times, with women making up the
majority of tea-pluckers, because their fingers are ostensibly more "nimble,"
and because they are perceived as being a more disciplined labor force.

Contemporary living conditions for tea workers have scarcely changed
from the nineteenth century; they are still housed in "line" rooms laid out in
perfect geometric rows on the estate, consisting of one or two rudimentary
rooms of one hundred square feet, with entire families crowded into a room,
in close proximity to the neighbors on either side. With a few exceptions, the
Tamil-origin laborers are not permitted to grow cereal or vegetable crops for
their own consumption, locking them into debt relations with the plantation.
Tamil workers were also denied electoral, property rights, or citizenship for
decades after independence, a systematic disenfranchisement which contributed significantly to the impetus for Tamil sectarianism, sowing the seeds for
the death-worlds and war ecologies of the civil war. Tamil laborers remain
substantially marginalized even today. Provision of education and medical
facilities is minimal, with the result that death rates on plantations are often
higher than in the rest of the country (Dawood 65); workers are subjected to
leeches, rashes, wasp and snake bites, pesticide poisoning, asthma from tea
dust, breast pain, sexual harassment, neck injuries from carrying tea bags
slung from their foreheads, and a lack of eating places, toilets, or facilities for

pregnant and nursing mothers. The Sri Lankan plantation is a "total institution," an agro-ecological system structured around the biopolitical control of labor and the bureaucratization and commodification of space and bodies (Duncan 67). The radical simplification of nature in which the land is rationalized into neat rows of tea shrubs, pruned to follow the slope of the hillside, and preventing access to other plantations, is mirrored by the rationalized lines of the dwelling places; both are intended to maximize extractive efficiency and surveillance of the isolated workers: "it was believed that smaller lines instilled a sense of belonging and decreased both fighting and plotting against the planter" (Duncan 84).

By the time of independence, even though the land devoted to tea, rubber, and coconut plantations remained relatively small in comparison to the country's total land area, Sri Lanka remained dependent on food imports purchased with the profits from export of tea and rubber, just as Ghana was tied to cocoa, Zanzibar to cloves, Guyana to sugar and rum, or Malaysia to tin and rubber. Sinhalese coconut kings and rubber barons dominated local politics, while Tamils were mostly consigned to plantation labor. The huge dependency on exports meant that local life remains at the mercy of an external market, where any fluctuation in commodity prices fixed by foreigners is "felt directly in the food baskets of the people" (Dawood 2), as in the present moment, where the conflict in the Middle East, a primary market for Sri Lankan tea, has led to a huge crash in exports, and food insecurity. During the colonial period, the surpluses derived from plantation were never reinvested in productive infrastructure, soil, or manufacture, but rather bled from the country back to the cores, as one British planter remarked: "Ceylon in the best coffee days was a sort of incubator to which capitalists sent their eggs to be hatched and whence some of them received from time to time an abundant brood leaving but the shell for our local portion" (Dawood 75). By the 1960s, four major producers controlled the majority of tea in the entire world: Brooke Bond Liebig, James Finlay, the House of Twining, and Lyons-Tetley; while a mere handful of people controlled the majority of plantation lands.

With the land reform of the 1970s, the state nationalized plantations, but did not take the opportunity to reinvest in their maintenance or adequately promote smallholder production and redistribution, and exports shrank as the bushes aged further, soil fertility declined, irrigation channels dried up, slopes eroded, and production withered. In 1992, when the estates were reprivatized as part of structural adjustment programmes enforced by the International Monetary Fund (IMF), some plantations were purchased by local companies, but most came under the control once more of large-scale TNCs, against which the small local capital class could not compete. In the twenty-first century, two producers now control the majority of the consuming tea markets: Unilever, the Dutch-British conglomerate, and Tata, the oligarchical Indian corporation which bought out Tetley. The agro-extractive regime of plantation remains little changed from the nineteenth century, and in some respects, Sri Lankan's domination by external forces and export-dependency is even more asymmetric, while increased competition from other producers drives down prices.

Ecological degradation in tea estates continues to accelerate, even more so than in rubber or coconut, with severe soil erosion and badly managed, aging trees. Planters frequently internalize the logic of boom-and-bust and adopt a speculative outlook, pursuing land-use policies that produce short-term profits to capitalize on the rise of prices in the volatile world market, rather than seeking small but steady yields over a longer period of time, based on long-term infrastructural investments in agricultural production and the practice of more sustainable farming. The rush to cultivate when the prices are high and the tendency to abandon estates or to cease maintenance when bushes age and prices crash only intensifies price volatility and the oscillation between boom and bust, since production costs rise at the same time as yields decrease and prices drop, lowering profit margins. The technicization of production with the green revolution which increased the scale of cultivation with high-yielding varieties and greater dependency on nitrogen fertilizers, has not mitigated the consequences of plantation monoculture, but only intensified the exhaustion of top soil fertility and the pollution of water-sheds and soils with chemical fertilizers (Talbot 96). The preference for "line planting" seedling plants in horizontal rows rather than contouring further aggravates mountain hydrology, precipitating large-scale erosion and land-slides. During the 1990s and 2000s, planters were fearful of investing in or maintaining their estates lest the civil war overspill and ruin their profits; calls to reinvest in the tea industry since 2010 have prompted some conversion of old estates to tourist attractions, renting out colonial bungalows, and turning factories into museums, but huge amounts of tea, coconut, and rubber land remain marginal or lie fallow. After a sustained run of periodic booms and busts across the twentieth century, the ecological regime of plantation in Sri Lanka seems near exhaustion or collapse.

COMMODITY AESTHETICS, PLANTATION MONOCULTURE, AND LITERATURE

In a seminal article, historian Sylvia Wynter has argued that the rise of the capitalist world-economy, whose accumulation regime emerged through the imposition of plantation-societies in the Caribbean, was a "change of such world-historical magnitude that we [in the Caribbean] are all, without exception, still 'enchanted,' imprisoned, deformed, and schizophrenic in its bewitched reality" (95). She argues that within Caribbean plantation monocultures, where people lack autonomous control over the local production of nature, history often seems like a "fiction written, dominated, controlled by forces external to itself," dictated by world-market prices or the decisions of colonial regimes or transnational corporations. The Sri Lankan historian Nawaz Dawood similarly imagines plantation in a series of catachrestic, irrealist metaphors, as like a "dye" introduced into water, where "everyone . . . was directly or indirectly affected by its introduction" (9), or as a debilitating parasite which sucks out the life-force from human labor and extra-human nature, exporting the surplus abroad:

> The plantation system produced in Sri Lanka, a social system with a distinct class structure, economy, and a way of life that separates it from other types of societies which are not plantation dominated. The impact of the plantation system on our society can also be compared to that of a parasitic plant on a tree; the tree and the soil work incessantly to feed it and whilst it grows, the condition of the tree itself becomes debilitated. Whether one is directly connected with the plantation system or not, is not material in such a society, for all members of such society are drawn into its vortex. (Dawood 2)

This imagery powerfully encapsulates the rupture in the social metabolism of nature which characterizes plantation monocultures, and is also suggestive of the situation of Sri Lankan aesthetics, in which plantation is ubiquitous but often paradoxically invisible, unremarked upon, read as background or as a landscape, rather than as the structuring principle around which a whole culture, with all its class relations and environments and forms of cultural production, is organized.

Michel-Rolph Trouillot has observed that the sugar export economy in colonial Haiti was not only the primary source of revenues, but produced a

> social culture: the socially drawn monopoly to subject to its refraction all other commodities and human beings themselves. Socially selected, socially identified, it became the principle around which human life was organized. Towns were built because of its proximity. Time was marked by its harvest. Status was linked to its possession. (372)

Likewise, tea, rubber, and coconut in Sri Lanka have their own social cultures organized around monopoly, producing rail and road networks, hill stations, plantation houses, institutions, and affects, particularly in boom periods, when excessive consumption by the planter elites and mercantilist classes ran high and a certain delirium characterized relations, as captured in Ondaatje's descriptions of his liquor-addicted family in *Running in the Family* (1982), or Romesh Gunesekera's descriptions of the coconut kings and rum barons in his novel *The Sandglass* (1998). However, it is not the hallucinations of the boom periods that are most often expressed in contemporary literature, but rather the exhaustion of the busts. The aesthetics of what I call Sri Lankan "plantation gothic" express a structure of feeling whose affects encompass the anxiety of the schizoid logic and price-volatility of the world-market, the dread of nature-society breaking down with the collapse of monoculture, the fear of nature's "revenge" in the form of blight, the nostalgia of the planter class for the booms. One of the most prevalent images in plantation gothic, especially in texts whose narration centers around the life-world of the Sri Lankan planter class, is that of the *waluwe*, the plantation estate house, usually abandoned, decaying, or in a state of spectral ruin. Depending on the subjective focus of the text, the *waluwe* can function as either critique or nostalgia, shifting in relation to different narratorial viewpoints of the changing class relations and political contexts of plantation in Sri Lanka.

In early incarnations, such as in Punyakante Wijenaike's *Giraya* (1971), the *waluwe* is tied to the politics of land reform and economic modernization in post-independence Sri Lanka. The delirious, hallucinatory aesthetics of the novel crystallize one of the most powerful examples of plantation gothic tied to rejection of plantocracy and commodity monoculture. On the eve of land reform, a young small-holder's daughter marries into a family which owns a decaying rubber and coconut estate, the descendants of a feudal *Mudaliyar* who grew rich during the earlier booms. Struggling to retain their feudal aristocratic privilege, the family are portrayed as frozen in a state of arrested development and bizarrely psycho-sexually warped by their dependence on a few commodities and extraction of surplus from their laborers, to whom they flaunt their alleged superiority. The estate is in a state of decay—with declining prices and production, they have been forced to sell off land and the plantation has shrunk to a hundred and fifty acres. A new road, intended to connect a new mill-factory to the city, will soon bisect their property. The prospect of industrialization and proletarianization of the laborers fills the family with dread, but is invested with all the optimism of early independence by the narrator, who describes the textile mill in organic terms as if it were alive, emphasizing that it represents the potential inauguration of a new ecological regime, surpassing plantation:

> Out in the village the textile mill was struggling to be born. While the after-noon slumbers, the mill grew spreading its roots like a giant banyan tree. [. . .]
> A young village maid passing by with a water pot heavy upon her hip told me that the mill would give employment to thousands of people. [. . .] She said proudly it would have three sections of workers: spinners, weavers, and others who would add finished touches. [. . .] I knew suddenly that this girl, the priest, and I felt the same way. Yes, life meant change, progress. The peace of the country was shattered by bulldozers at work, cranes, cement mixers, and swarms of women. We liked this. [. . .] But I knew that should this noise

reach the ears of those slumbering within the old walauwe it would not be welcome. (Wijenaike 31)

By contrast, the plantation house is described with paranoiac anxiety as a kind of decaying body, metonymic of the collapse of the plantation regime and of the privileged social relations attached to it:

> In daylight it looks worn and crumbling. The white trellis work below the gutters has rotted away and only stumps remain like old decayed teeth. The gutters themselves are rusty and full of holes so that they turn into spouts during the rainy season. [. . .] The rounded pillars that support the roof over the porch bears signs of age and neglect. [. . .] The family crest stamped on the front of the porch is covered with a fungus. (17)

In the gardens, "mosquito larvae wriggle freely" (18) and weeds fill the flower beds; on the estate, palms are untended and huts unthatched, while the workers believe that "on plantations ghosts walk freely" (44). At the novel's conclusion, the narrator is freed from the oppression of the house and family, when her mother-in-law is slain by her jealous servant, and her husband is revealed as an illegitimate son with no title to the property, so that the estate descends into the hands of the superintendent-laborer instead. With this revelation, the haze of sexual fetishism that permeates the narrator's consciousness as metaphor for the larger social fetishism of the commodity monoculture lifts and dissipates, freeing her to pursue a newly autonomous life as a career worker in the capital. Her escape from the plantation thus coincides allegorically with post-independence land reform and the prospects of liberation from the asymmetries of feudal and colonial relations. The gothic revelation of the repressed past performs a kind of exorcism that enables the imagination of a national future beyond plantation and plantocracy, albeit bound to a telos of economic modernization which fails to predict the social consequences of liberalization.

When the plantation house trope returns in later literature, it expresses not futurity, but rather the failure of the post-independence prospects of liberation. Literature of the 1990s and 2000s draws on plantation gothic aesthetics to express a sense of double haunting by the ghosts of both the civil war and continued plantation. Far from being eliminated by the modernization programmes of the earlier state, the infrastructure, class relations, spatial organization, and ecological simplification corresponding to the plantation regime persist into the present, even as the regime is portrayed as being even more unstable and vulnerable to exhaustion. Thus, in Roma Tearne's *Mosquito* (2007), the artist Theo lives in a moth-eaten, beetle-drilled estate house, stripped of its original purpose but redolent of colonial nostalgia: "It was a useless house really, everything was broken or badly mended, everything was covered in fine sea sand. [. . .] Objectively, it might have made a better relic than a house, but relics were plentiful and houses of this size not easily found. The fact was Sumaner House was huge. Once it must have been splendid" (Tearne 43). The former estate lands lay fallow and waste, haunted by rumors of ghosts:

> On the edge of Aida Grove, on a slight incline not far from Sumaner House was an area where coconut trees would not grow. The earth was bared and wasted, without grass, without bushes, without life. There was nothing there except a line straggling tamarind tree. Once this patch of land had been part of a larger grove of coconuts. It had belonged to the owners of Sumaner House, but with time, neglect, and some erosion it had become common land, useless and uncultivated. Superstition abounded. (Tearne 107)

Later in the novel, Aida Grove becomes the site of a lynching attached to the civil war, and the eastern province is described as reduced to a wasteland by the intrusions of the war: "Once, this had been a fertile land with rice and coconut as the economy. Once, this had been a tourist paradise, lined with rest houses. The port had been an important naval base. Now there was not a single person in sight" (Tearne 199). The war ecology inscribes territories already spatialized by plantation in ways which increase vulnerability, as when the abandoned palm groves become breeding grounds for mosquitoes, provoking a malaria outbreak across the island: *"Further out towards the coast, the rainwater filled the upturned coconut shells, as they lay scattered across the groves. Here the beautiful female anopheles mosquitoes, graceful wings glinting in the sun, landed lightly and prepared to create a canoe of death for their cargo of eggs"* (emphasis original, Tearne 9). Jason Moore argues that disease vectors and epidemiological events frequently proliferate at the collapse of an ecological regime, and throughout the novel, malaria functions both as material sign of a real historical phenomenon—the resurgence of malaria due to the inability of the government to spray pesticides during the civil war and to the accumulation of stagnant water across vast tracts of abandoned estates and jungle—and as a literary metaphor of ecological exhaustion exacerbated by political violence. However, the trope's ecological valence is compromised by its subsequent entanglement with inflammatory racial discourses, when Tearne compares LTTE female suicide bombers to anopheles mosquitoes at the conclusion of the novel: "a new breed of women from the north of the island . . . these women were the new trailblazers, the world epidemic slipping in unnoticed, just as the malaria season returned" (253) constructing the bombers' actions in racialized epidemiological terms, rather than understanding them in political terms as emerging from particular socio-economic conditions. Plantation collapse and war ecology are indelibly linked in the text, but with a strong ecophobic content that fails to fully interrogate their imbrication.

The nostalgic portrayal of the crumbling plantation house in Tearne's novel is representative of a general trend in much of Anglophone Sri Lankan fiction aimed at an Anglo-European market, which tends to privilege a middle-class view or to focus on the history of local planter elites. In contemporary uses of plantation gothic, the declining plantation most often serves not as economic commentary, as in the earlier *Giraya*, but rather as anxious metonymy of a class whose fortunes are threatened both economically and politically by the ongoing ethnic violence. The actual labor of Tamil-origin workers on plantation estates, or the everyday business of commodity extraction and production of nature is often far less visible in fiction than the railways, roads, hill stations, estate houses, and bourgeois class relations which it has produced. A characteristic example is the chapter "Tea Country" in Michael Ondaatje's lyrical memoir *Running in the Family*, which is permeated with a complicated nostalgia for his Burgher ancestors' drunken exploits as plantation overseers at the height of the early twentieth century boom. In the limpid prose of his description, the tea country of the hills emerges in a pastoral, idyllic light as a "sleepy green landscape," "the green pattern of landscape and life-style almost unchanged" (Ondaatje 165). As he gazes out from the tea bungalow, the radical simplification of nature is applauded rather than rued, in a passage which is resonant of colonialist perceptions of jungle made civilized, while tea workers appear only once, in a passing reference that emphasizes their silent, subaltern homogeneity: "I can leave this table, walk ten yards out of the house, and surrounded by versions of green. The most regal green being the tea bush which is regal also in its symmetrical efficient planting. Such precision would be jungle in five years if left alone. In the distance the tea pickers move, in another silence, like an army" (Ondaatje 167).

Similarly, in Ameena Hussein's novel *The Moon in the Water* (2009), the tea country is described in uncomplicated, pastoral terms: "The countryside wore a dress of green and white—tea bushes and waterfalls. A delicate sweep of rolling hills that gave way to the great rainforest Sinharaja beyond" (9). When the protagonist Khadeeja visits a low country tea estate, the narration emphasizes its decline from earlier days:

> It had been sold and bought back as the family fortunes dwindled or grew. Mostly sold and rarely bought back, the once majestic, over five hundred acre state dropped through many generations and was now reduced to a shameful twenty. [. . .] Man and Nature had both taken a toll on the once imposing estate which now lay in three pieces like a fragmented Grecian urn. [. . .] The house, a badly renovated walauwa, sat on top of a very small hill. It overlooked all the lands it owners once possessed, from the paddy fields that fell away below to the neighbouring hills and hillocks that rose up around it. (59)

Unable to turn a profit, the owner, Arjuna, is forced to diversify the estate's production, converting the house into a guesthouse, and the plantation to eco-friendly practices: "It was no longer a tea estate exclusively; in fact no-one quite knew what kind of estate it had become. It was experimental with a little forest and a medium sized ayurvedic plantation on one side; with a fruit and vegetable strip and a timber plantation on the other" (59). Arjuna feels shame that he does not own the estate, that it was bought from his family with his German wife's money, bitterly describing himself as a "a *hamu murakaraya*, a gentleman watcher," a "kept man [. . .] living the good old *walauwa* life," "a true blue Arya Sinhala man" (90), rather than the *Hamu Mahattaya*, the plantation lord of yesteryear which he longs to be. The novel focuses on his melancholia at the perception of his parasitic relation, and his anxiety at being in thrall to foreign capital—as allegorized by his wife—rather than exploring the workings of the estate. As with Ondaatje's memoir, the single reference to conditions of labor is when Khadeeja pretends to be a "social anthropologist studying the housing conditions of the tea-pluckers" but promises she will not be conducting new research, thus eliciting his relieved reaction: "it's a good thing that my tea-pluckers will not be harassed by you" (64).

Jean Arasanayagam's poetry echoes the tropes of nostalgia for the culture of the colonial boom periods, offering multiple images of abandoned estates, as in "Peacock Country," with its elegy for ancient, "wide-trellised" veran-dahs and "abandoned bungalows," emptied of their "imperial owners," their wooden shutters askew in the wind (*Searching* 38). The uncanny emptiness of evacuated estates is paired with anxiety at the ease with which the un-tended plantation ecology gives way to a less regimented form of nature, recalling Ondaatje's observation that the plantation's "precision" would easi-ly give way to jungle. While the estate house's "thick-girthed pillars" still shoulder the traces of "individual families" and "forgotten epics," distorted, gnarled forms of the plants and trees threaten to run wild and overtake the once manicured lawns, growing in "sculptured contortions" with "sparse unkempt leaves" throughout the "untended gardens and neglected pathways" (*Searching* 38). However, the rainforest and tea country are more polyse-mous in Arasanayagam's work than in Ondaatje's or Hussein's, functioning alternately as sacral space, as forest-garden meaningfully shaped by human labor, as a dangerous site of terrorist activity, or as a teeming site of untamed fecundity, both exhilarating and threatening. Her work oscillates between the deployment of pastoral images that express a longing for a rootedness in land and nature which might form the basis of a green poetics, and the attempt to negotiate or repudiate the territorializing discourses which traditionally ima-gine the jungle in terms of anthropocentric ownership or economic rational-ization.

Significantly, Arasanayagam's short fiction offers one of the few correctives to the paradox by which plantation and estate houses are ubiquitous in contemporary Sri Lankan fiction, yet plantation labor absent. In her short story "The Journey," the Sinhalese refugee narrator compares his own flight to the historical death-marches of the Tamil indentured workers in the nineteenth century:

> I am reminded of the stories about the plantation workers who were brought to our island two hundred years ago. Brought from South India in their hundreds in ships. Disembarking at Talaimannar, they made the long trek from the north, through thick, animal-infested jungles, to the central highlands to work on the tea estates. So many died on the way of cholera, dysentery, malaria [. . .] leaving their skeletons as new landmarks on that terrifying journey. And of those who reached the central highlands, many hundreds died of fever, chills, pneumonia in those mist-veiled mountains. (*All* 4)

The narrator remarks that to the traffickers, "we are not human beings [. . .] we are 'dollars'" (*All* 4). So, too, were the Tamil workers conceived not as people, but as money, within the plantation monoculture: humans reified as "hands," valuable only insofar as they create surplus value. In other stories, she similarly subjectizes the experiences of workers on prawn aquaculture farms and coconut plantations, traces the effects of paradise tourism, and narrates the experiences of Sri Lankan migrant workers in Kuwait.

The fiction of Romesh Gunesekera, however, offers the most complex view of plantation ecology. In *Heaven's Edge* (2002), a novel set in a post-apocalyptic future on an unnamed island resembling Sri Lanka, the narrator Marc encounters an abandoned tea estate, converted first into a museum, then abandoned once more after an island-wide civil war. The passage repeats the usual tropes of spectral ruin and evacuation, but is distinctive in how it moves beyond nostalgia, aiming instead to recall the sensations and lived experience of labor in the factory:

> I made my way to the main factory floor. The place was gutted. All the machines had been removed, but the interior still smelled of tea. It rose out of the floorboards and off the walls and seemed to stain the air with the odour of old ghosts. In its heyday who would have been here? Sometimes it is so difficult to remember who belongs where, when, or why. Whose was the labour, and whose the capital? There would have been blasts of hot air and the noise of dryers and rollers; wheels turning, the smell of burning, roasting tea. Narcotic sweat. There was a time when the sound of machines would have filled the air all around the hills. Factories in full swing. A steam train chugging up to the central hill towns. Eldon loved to recall those scenes, complete with sound effects: the clacking of wheels, the hoot of the engine, the constant gabble of conversations between strangers. (119)

Rejecting his grandfather's halcyon memories of the tea boom as "sanctimonious claptrap," Marc seeks to unearth the precise socio-ecological structures which underlay the plantation—who worked there, who invested in it. He strives to re-materialize the ghostly relations of the past, rather than mourn the demise of a golden age. Similarly, he is not threatened when he encounters both tea and coconut plantations which are moving from a rationalized to a less regimented state of nature, instead seeing signs of liberation in the unfurling vegetation: "the once tightly curled tea bushes, slackened with neglect, seemed to be stretching out for freedom" (123). On the abandoned coconut estate, he harvests nuts from the "young, wilder trees" and turns the "whole plantation into a self-sustaining refuge," a forest garden, learning to "use coconut husks for moisture retention, fiber as mulch, recycle waste through organic compost-generation" (192–93). The novel offers a sustained critique of plantation monoculture and the export dependency and degraded ecologies it produces—the "overexploited cashlands . . . and leached scrubland that retreating global markets and destitute governments left in their wake" (36)—while attempting to visualize a future in which plantation could be superseded as the primary mode of organization of socio-ecological relations, even if this own vision is cut short by the novel's own conclusion, when Marc's refuge is tragically invaded by military forces, as if to suggest the difficulty of actually bringing an alternative future into being.

The selection of texts examined in this chapter are meant to be suggestive, rather than comprehensive, indicating the variable spectrum of literary registrations of Sri Lankan commodity frontiers, in which plantation features both as thematic content and as aesthetic form, sometimes featuring only as background or environmental unconscious, at other times serving as the organizing principle of a text, and taking on different resonances according to the historical period, the phase of boom-bust, and the class perspective of the narration and the subjective world of the text. I have attempted to sketch the lineaments of a mode of world-ecological analysis, through which the environmental content of Sri Lankan texts might be read as registering the material relations of the ecological regimes which have powerfully structured the island's history and culture and enforced its subordinate relation to a world market, concentrating on several textual encodings of tea, rubber, and coconut monocultures as examples of the huge terrain which remains to be explored and the many possibilities for comparison with other commodity frontiers across the Global South.

WORKS CITED

Arasanayagam, Jean. *All is Burning.* New Delhi: Penguin, 1995. Print.

———. *Fault Lines: Three Plays.* Colombo: Godage International Publishers, 2008. Print.

———. *Reddened Water Flows Clear.* London: Forest Books, 1991. Print.

――――. *Searching for an Ambalama: Poems.* Colombo: Godage International Publishers, 2009. Print.

Dawood, Nawaz. *Tea and Poverty: Plantations and the Political Economy of Sri Lanka.* Kowloon, Hong Kong: Urban Rural Mission, 1980. Print.

Dharwadker, Vinay. "Poetry of the Indian Subcontinent." *A Companion to Twentieth-Century Poetry.* Ed. Neil Roberts. Oxford: Blackwell Publishing, 2001. 264–80. Print.

Duncan, James S. *In the Shadows of the Tropics: Climate, Race and Biopower in Nineteenth-Century Ceylon.* Aldershot, Hampshire: Ashgate, 2007. Print.

Gunesekera, Romesh. *The Sandglass.* London: Granta, 1998. Print.

――――. *Heaven's Edge.* London: Bloomsbury, 2002. Print.

Hussein, Ameena. *The Moon in the Water.* Colombo: Perera Hussein Publishing House, 2009. Print.

Moore, Jason W. "The End of the Road? Agricultural Revolutions in the Capitalist World-Ecology, 1450–2010." *Journal of Agrarian Change* 10.3 (2010): 389–413. Print.

――――. "Wall Street is a Way of Organizing Nature." *Upping the Anti* 12 (2011): 47–61. Print.

Niblett, Michael. "World-Ecology, World-Economy, World Literature." *Green Letters: Studies in Ecocriticism* 16 (2012): 15–30. Print.

Ondaatje, Michael. *Running in the Family.* London: Picador, 1984. Print.

Peebles, Patrick. "Profits from Arrack Renting in Nineteenth-Century Sri Lanka." *Modern Sri Lanka Studies* 1.1 (1986): 65–83. Print.

Talbot, John M. "Tropical Commodity Chains, Forward Integration Strategies and International Inequality: Coffee, Cocoa, and Tea." *Review of International Political Economy* 9.4 (2002): 701–734. Print.

Tearne, Roma. *Mosquito.* London: Harper Perennial, 2007. Print.

Trouillot, Michel-Rolph. "Motion in the System: Coffee, Color, and Slavery in Eighteenth-Century Saint Domingue." *Review* 3 (1982): 331–388. Print.

Webb, James. *Tropical Pioneers: Human Agency and Ecological Change in the Highlands of Sri Lanka.* New Delhi: Oxford UP, 2002. Print.

Wynter, Sylvia. "Novel and History, Plot and Plantation. " *Savacou* 5 (1971): 95–102. Print.

Wijenaike, Punyakante. *Giraya.* Panaluwa, Padukka: State Printing Corporation, 2006. Print.

Chapter Three

Scenes from the Global South in China: Zheng Xiaoqiong's Poetic Agency for Labor and Environmental Justice

Zhou Xiaojing

In the early-morning hours of July 16, 2009, Sun Danyong, a twenty-five-year-old employee of Foxconn, the largest electronics manufacturer in the world, committed suicide by jumping from the twelfth floor of an apartment building on a factory campus in Shenzhen, the premier economic development zone in Guangdong Province, Southern China. This incident marked the beginning of a series of "Foxconn suicides," which occurred between January and November 2010, when eighteen Foxconn employees attempted suicide, with fourteen deaths.[1] Despite the subsequent media attention and investigations by *New York Times* and by the Fair Labor Association at the behest of Apple, one of Foxconn's clients, Foxconn workers' suicides continued into 2012.[2] Around 8:00 a.m. on March 17, 2012, Tian Yu, a seventeen-year-old female worker jumped from her fourth floor dormitory at Foxconn's giant Longhua plant, home to more than two hundred thousand workers in Shenzhen. According to a news report by Mimi Lau, which appeared in the *South China Morning Post*, Ms. Tian was "the fourth Foxconn worker across the country to attempt suicide this year, the youngest, and one of only four survivors. Up to last month, fourteen others, aged from eighteen to twenty-five, had died" (Lau, Web). Ms. Lau is one of the approximately nine hundred thousand workers in Foxconn's twelve factories located in nine cities in China. About five hundred thousand workers are employed in its major Shenzhen plants (Lau, Web). Like the other Foxconn factories, the Longhua Science and Technology Park, a.k.a. "Foxconn City" or "iPod City," which covers about 1.16 square miles, is a walled campus with fifteen factories, dormitory buildings for the workers, a swimming pool, a fire brigade, its own television network (Foxconn TV), and a city center with a grocery store, bank, restaurants, bookstore, and hospital. About one quarter of the employees live in the dormitories, and many of them work up to twelve hours a day, six days a week. These armies of cheap labor are part of an emergent "Global South" in China, resulting from economic globalization and local development.

Referred to as a "sweatshop" corporation by Chinese workers, Foxconn Technology Group is a subsidiary of the Hon Hai Precision Industries Ltd., which is 112th in the ranking of Global Fortune 500 companies. A report entitled "Workers as Machines: Military Management in Foxconn," by the advocacy organization Students and Scholars Against Corporate Misbehaviour (SACOM), notes that according to "market research firm iSuppli Corp., in 2009, Foxconn took over 44 percent of the global revenue of the entire electronics manufacturing and services industry." According to an iSuppli estimate in 2010, Foxconn would gain half of the industry revenue by 2011.[3] And it is going to expand the workforce to 1.3 million people by the end of 2011.[4] It seems that low wages and excessive overtime, in part, underlie Foxconn's large profits, as well as its workers' suicides.

While the suicides of Mr. Sun and Ms. Tian were acts of hopelessness and silent protests against unfair treatment by Foxconn, other workers used suicide as a strategy for demanding fairness and rights. On January 2, 2012, when the Foxconn managers decided to move six hundred workers to a new production line for making computer cases for Acer, a Taiwanese computer company, a group of a hundred and fifty workers threatened to commit suicide from the top of their three-floor plant in Wuhan, a city in Southern China.[5] "We were put to work without any training, and paid piecemeal," said one of the protesting workers, who asked not to be named. He added: "The assembly line ran very fast and after just one morning we all had blisters and the skin on our hands was black. The factory was also really choked with dust and no one could bear it" (Moore, "'Mass Suicide' Protest"). According to Malcolm Moore, a reporter for *The Telegraph* stationed in Shanghai, Foxconn factories are run in such a "'military' fashion that many workers cannot cope. At Foxconn's flagship plant in Longhua, 5 percent of its workers, or 24,000 people, quit every month" ("'Mass Suicide' Protest"). The instability of Foxconn's workforce and "Foxconn suicides" in China reflect more than the dehumanizing working conditions and living environments for those who are reduced to machine-like cheap labor. It demonstrates the workers' resistance to subjugation and exploitation by a new form of "empire" emerging from a postcolonial, postmodern world system of transnational corporations in the age of economic globalization. While Foxconn's multinational systematic operations of production for maximizing profit evoke colonialism, the unprecedented scale of such operations with transnational mobility in the accumulation of capital has led to profound social and environmental transformations in China, including mass migration of laborers from the countryside to cities. In fact, Foxconn's walled-in "sweatshop" factory-campuses and other emergent landscapes such as "villages within the city"—new urban neighborhoods where migrant workers live, and where poverty and crime rates are high—constitute a unique geography of a Global South in China.

The poems of Zheng Xiaoqiong, who used to be a migrant worker in Southern China, gives voice to this emergent Global South in China. Born in 1980 in rural Nanchong, a town in Sichuan Province in Southwestern China, Zheng left her home village in 2001 for Dongguan City in southern Guangdong Province, where she became a migrant worker for seven years, moving from factory to factory, including a hardware plant, a plastic factory, a garment factory, and a furniture factory. While recovering from an injury at work in the hardware plant, she began to write poems, which appeared in various prestigious literary magazines, including *Chinese Poets, Bell Mountain*, and *People's Literature*, and in anthologies such as *The Best Poetry of the Year* and *The Best Prose Poems of the Year*. She has published several collections of poems, including *Huang Maling Village* (2006), *The Depths of Night* (2006), *Two Villages* (2007), *Selected Poems by Zheng Xiaoqiong* (2008), *Dark Night* (2008), *Pedestrian Overpass* (2009), *Poems Scattered on the Machine* (2009), *Aches and Pains* (2009), *Pure Plants* (2011), and a collection of poems and prose titled *Female Migrant Workers* (2012).

Zheng's writings have won numerous literary awards, including the Lu Xun Literary Award of Guangdong Province, the Dongguan Lotus Flower Literary Award, the 2006 Chinese Avant-garde Literary Award, and the Liqun Literature Award from *People's Literature* in 2007, among others.[6] Recognized as a representative voice of the contemporary age, Zheng's poetry has drawn enthusiastic attention from the literary establishment and the Chinese media. She was invited to become a member of the Guangdong Writers' Association and the Literary Academy of Dongguan. She was also named by the journal *Poetry* as one of "the ten best young poets of the new era" (*Poetry* November, 2009). With her background and her advocacy for social and environmental justice, particularly for migrant workers, Zheng stands out among her fellow distinguished poets. She was selected in 2007 by the widely circulated magazine *Chinese Women* as one of the ten most influential figures of the times at home and abroad. In 2009, she was elected as a delegate from Guangdong to the Eleventh People's Congress. Zheng was hired as an editor of the literary magazine of the province based on her literary achievements and her training in a writing workshop for migrant workers given by a magazine publisher in Guangzhou, the capital city of Guangdong.[7] Despite all this renown, Zheng refuses to become a representative of the "success story" of millions of migrant workers, who continue to live at "the bottom" of society. While a large number of her poems focus on migrant workers, the recurring central concerns in her works include issues of pollution, environmental destruction, and ecological degradation.

UBIQUITOUS, PENETRATING IRON AND TOXINS

Critics in China have offered insightful readings of Zheng's poetry. Zhang Qinghua in his essay "Who Touches the Iron of the Age: On Zheng Xiao-qiong's Poetry" notes the power of Zheng's unique diction and imagery, which not only reveal the marginalized lives at the lower strata of society, but also capture the characteristics of the contemporary era through a "new aesthetics of iron":

> "Iron" 铁, the word with a primitive root itself, is the symbol and center of her poetry, as this Chinese character appears frequently throughout Zheng's poems. As a symbol of cold, hard, industrialized existence, as a metaphor of mass production assembly lines, as an alien power in sharp contrast to our vulnerable human bodies and nature, iron expresses "the aesthetics of [the] industrial era" and can be said to have an irreplaceable significance. Iron is the darkness and order as well as the soul and destiny of the age. It rules over this world as the humble life of flesh and blood appears powerless to resist it. If Zheng's poetry has a unique aesthetic importance, its most salient metaphoric extension would be the way she offers to our age the cold, hard, new aesthetics of iron. (35)

While Zhang focuses on aesthetic significance of Zheng's poetry, Gong Hao-min points out its ecological concerns in his essay "Toward a New Leftist Ecocriticism in Postsocialist China: Reading the 'Poetry of Migrant Work-ers' as Ecopoetry." Gong discusses the characteristics of migrant workers' poetry which indicates the manner in which the poets' sense of their identity shapes the ways they "conceive their relationship with nature" (148). Apart from "bearing witness to the pollution and the destruction of nature in the unchecked process of industrialization and urbanization, these poets, dis-oriented in an urban environment, usually see their own experience in paral-lel with a nature disturbed by human interventions. For instance, plants up-rooted from the earth are an image that appears repeatedly" (148). Moreover, Gong contends: "Migrant worker poets record, from various angles, how this monstrous machine called global capitalism engulfs their body and soul, and the natural environment around them" (151).

Although both Zhang and Gong's insights shed light on some significant aspects of Zheng's poetry, they overlook other related issues that Zheng confronts, such as vocational diseases, disintegration of migrant workers' families, and the "hollowing" of villages, as well as migrant workers' resistance to subjugation and exploitation. My reading of Zheng's poems seeks to foreground the emergent Global South China and its protests, resistance, and challenges from the migrant workers' perspectives.

In an interview about migrant workers and Foxconn following the Foxconn suicides, Zheng talks about the workers' plight. She notes that an overwhelming, inescapable feeling she and others like her experience is fatigue— a lingering weariness of both body and spirit.[8] In her poems, Zheng seeks to bring to light the condition of migrant workers, especially female migrant workers. Her poem "Weariness" captures both the physical and psychological dimensions of their experience:

> She is weary of untimely youth, this familiar scene . . . .
> . . . .
> Lying exhausted, she is weary of time-cards, machines
> Ah, what time is it?
> she is still standing by the window, the scorching blue sky
> has floating clouds, mirages of her future, love in a picture, and
> Childhood buried in distant memories, she'll still write down
> life in a strange town on the back of her time cards,
> remember her family far away
> Beyond her overtime hours, meager pay, vocational diseases
> She is aging by the side of machines, on workshop decks, construction sites
> Behind her, cities of high rises one after another
> Have already deserted her. [9] (*Selected Poems* 49)

Along with the feeling of exhaustion that permeates the speaker's mind, body, and environment are feelings of futility, loss, marginalization, and isolation.[10] Yet an indomitable spirit prevails in the speaker's articulation of agency through her writing. She expresses the determination to write about her life as an exploited, displaced, neglected migrant worker in the alienating city. Writing, then, for the migrant-worker speaker in the poem, as for Zheng herself, is a form of resistance—resistance to becoming invisible, silent, marginal, and dispensable while at the same time being exploited as cheap labor indispensable for industrial prosperity.

For Zheng, writing about migrant workers' experience is an ethical question that is inseparable from aesthetic concerns and manipulation with language and form. She explores the possibilities of enacting the experience of migrant workers through manipulation of language and arrangement of lines on the page in poems such as "Migrant Labor, Words of Vicissitudes":

> Writing down this phrase "migrant labor" is difficult
> Utter it with tears running when in the village
> I thought it a ladder for sending my life soaring once again yet upon arrival here
> I find it a trap an injured finger
> . . . .
> Migrant labor is a label it marks you for sale on the market
> Migrant labor allows you to be fed in others' troughs

Migrant labor makes you wander all year round

. . . .

You must sell 300 *jin* of rice for a bus ticket to get out of your village
400 *jin* of wheat for a temporary residential card, health card, family-planning card
Unmarried card migrants' registration card work ID card border-zone card
You age and languish under the weight of these cards
I live forever in the language of migrant labor making home
In a floating shoe. . . . (*Pedestrian Overpass* 71–73)

The broken lines and disrupted syntax mirror the dislocation, uncertainty, and fragmented lives of migrant workers. As the speaker states by the end of the poem, while seeking to redeem and reclaim migrant workers' trampled lives, she simply "can no longer calmly, quietly, silently write poems / when saying once again 'migrant labor' . . ." (74). For this phrase "migrant labor" "no longer inhabits a poetic landscape. / To live with this phrase, one must endure unemployment, begging to be saved / running around expelled insomnia and security officers / looting in the name intended otherwise . . ." ("Migrant Labor" *Pedestrian Overpass* 74).

In searching to find a suitable language for portraying migrant workers' experience, Zheng has developed a poetics that characterizes the ways in which the workers' physical and psychological conditions are shaped by their working and living environments, which are produced by both local and global forces. Entangled social, cultural, and physical environments play an active role in giving rise to the phenomenon of migrant workers, and in altering their physical, psychological, and social statuses. The concept of "trans-corporeality" proposed by the feminist ecocritic Stacy Alaimo offers a theoretical framework for understanding the multiple interrelations and effects underlying Zheng's poems. Alaimo defines trans-corporeality as "the time-space where human corporeality, in all its material fleshiness, is inseparable from 'nature' or 'environment'." She contends:

> Trans-corporeality, as a theoretical site, is a place where corporeal theories and environmental theories meet and mingle in productive ways. Furthermore, the movement across human corporeality and nonhuman nature necessitates rich, complex modes of analysis that travel through the entangled territories of material and discursive, natural and cultural, biological and textual. ("Trans-Corporeal Feminisms" 238)

Such movement across the body of the migrant worker and its environments in intertwined, heterogeneous networks characterize Zheng's portrayal of migrant workers' situations. Her poems such as "An Iron Nail," "Iron," "Nails," and "Lungs" suggest that the materiality of an industrial environment has penetrated the workers' bodies and minds, shaping their sense of self as well as the prospects of their future. In "An Iron Nail," Zheng employs an iron nail as an analogy for the life of a female migrant worker:

In the furnace, she has been cast an iron nail
Now being placed on the wall, she feels lonely and her lower half
Cold; her master hangs on her
Plastic bags that hold vegetables, onions, eggs and
Fatty pieces of meat; she approaches life in silence
Time is rusting on her body, her lustrous youth
Turning withered yellow, dark red; she does not speak, even though
Memories constantly bring her back into her once fiery past
She could have become a machine in operation, but being inattentive
She's turned into a small iron nail, like an invisible person
Standing on the wall, witnessing mediocre, trivial lives, and yet
Her inner life is so utterly banal, her outlook is not
Completely hopeless; on the contrary, she loves the ordinariness
Of everyday living, the endless quarrels and babbling chatters of
Her master; she's resigned to the daily tranquility, half of her body
Sunk in the wall, time is accumulating gradually as
Rust is swallowing her little by little; every day, she endures
Taunting from shiny iron tools, the great God has molded her
Into an iron nail; her life is already a failure
Now fixed on the wall, she is doubly unfortunate
But she never complains or resents, she has forgivingly accepted
Her fate; for she knows being alive is more courageous and peaceful than being dead.
(*Poems Scattered on the Machine* 15–16)

Although the female migrant worker who has turned into an iron nail does not complain, the poem itself is a powerful protest against the injustice of dehumanizing exploitation through layered meanings embedded in the images. The barren, desolate, and hopeless life of the female migrant worker metaphorized as a nail fixed on the wall serving a diminutive yet useful, subordinate function for her master's daily living contrasts his well-to-do yet banal commonplace life, which depends on the nail for a particular purpose. The critical distance created by the speaker's ironic tone also exposes the fact that the limited choices in life have conditioned the female migrant worker's outlook, symbolically crippling her for life. Thus the dominant image of the iron nail points to systemic conditions that confine and devalue the migrant workers' existence.

Zheng often situates the migrant workers' condition in the larger context of industrialization and globalization. The changing landscape, including urbanization and environmental destruction, in "The Lychee Grove" seen through the eyes of a migrant worker, at once reflects and critiques the complex impact of industrialization equated with prosperity, progress, and Westernization. The river that "still keeps / The slow pace and sadness of the past era," is likened to "a sick person" whose sickness is marked by industrial pollution—"It's greasy, swarthy, stagnate with the stench of industrial wastes" (*Poems Scattered* 35). Yet,

No one listens to its soft sobbing. Up on the hill

Excavators are digging up lychee trees, the felled trees
Lie on naked yellow dirt, their tiny flowers
Scattered on the ground, their faint fragrance fading, ah, in the setting sun
I see on the Phoenix Boulevard, so many like me
Who come from far away to share this flourishing prosperity of the industrial age
And to witness where the lychee trees are cut down factories and machines rise,
Chinese-style bluish-green tile-roofed houses are replaced by Western scenes
In this village, no one but me
Listens for what's behind booming economy—the sobbing cold stream and the
Sorrow of the felled lychee grove, the conservative village shrine amidst forests
of high rises. (*Poems Scattered* 35)

Like migrant workers who are suffering from industrial diseases, the river is "sick" because of industrial pollution. Their plight is part of the social, cultural, and environmental transformations taking place. While the scenes of urbanization and industrialization are localized in this poem, they are connected to economic globalization in other poems.

In the poem "Iron," the speaker links the ubiquitous iron in the migrant workers' everyday lives to "an export product for the U.S.," highlighting the impact of transnational capital of the Global North on the lives of those who have been reduced to cheap labor (*Selected Poems* 81). Zheng pursues further this unequal relationship in her poem "Nails." Using nails as an extended metaphor, Zheng indicates the dehumanization of workers, who have become parts of machines for production and profit-making. Yet, unlike machines, countless workers suffer from endless fatigue, industrial maladies, violations of their rights, homesickness, and a sense of being trapped in a meaningless, hopeless situation as cheap labor. Thus, the "trans-corporeality" in this and Zheng's other poems about migrant workers entail more than "the material interconnections of human corporeality with the more-than-human world," which Alaimo's concept of "trans-corporeality" emphasizes ("Trans-Corporeal Feminisms" 238). While "Nails" demonstrates the inseparable connections between "environmental health, environmental justice, the traffic in toxins"—connections which the concept of trans-corporeality helps highlight ("Trans-Corporeal Feminisms" 239), the poem reveals the psychological and emotional dimensions of migrant workers' conditions of life and work, which are also shaped by economic globalization:

How much love, how much pain, how many iron nails
Have fixed me on the machine board, blue prints, order forms
Morning dew, midday blood
One iron nail holds together overtime, vocational diseases
Unnamable grief, pegging migrant workers' lives
On complex buildings to spread the fortunes and misfortunes of an era
How many fatigued shadows in the dim lights
How many numb smiles of fragile, thin, and small migrant-worker sisters
Their love and memories are like the moss under green shades, quiet and fragile

How many mute nails pass through their placid flesh and bones
Their youth imbued with kindness and innocence alongside profits, wages in arrears
Labor Law, nostalgia, and an incomprehensible love
On the pale blue assembly line, deck seats dangle
One after another, piercing nails pause a moment
Outside the window, autumn's passing, someone relies on nails to survive.
(*Selected Poems* 51)

The references to "order form," "profits," "overtime," "wages arrears," "unnamable grief," and "vocational diseases" at once allude to and protest against larger systems and practices of exploitation and environmental injustice beyond the site of production or national borders.

The implied connections between local and global, between working environments and migrant workers' work-related illnesses become more explicit in poems such as "She" and "The Age of Industry" collected in *Poems Scattered on the Machine* (2009). In "She," Zheng situates a female migrant worker's ten-year life in a deck seat on the assembly line and her physical ailments in an ecological web that links the U.S. consumers' market to migrant labor, deforestation, and economic development in China. While the fatigued, diseased, aging working-class female body in this poem testifies to what Alaimo calls "the penetrating physiological effects of class . . . oppression, demonstrating that the biological and the social cannot be considered separate spheres" (*Bodily Natures* 28), Zheng again links the conditions of this oppressed, exploited, working-class body to products on the market in the Global North:

Time opens its gigantic mouth, the moon is rusting
On the machine, haggard, dim, turbid; ominous danger in her heart
Flows like a turbulent river, the cliffs of her body collapse; dirt and gravel
Tattered pieces of time fill up the ferocious river of her female body
Confused tides no longer fluctuate according to season, she sits in a deck seat
As moving products interweave with time, devouring everything, so fast is she
Aging, ten years flew by like water . . . immense fatigue
Drifts in her mind . . . for so many years, she's been keeping company
These screws, one, two, turn, left, right
She fixes her dreams and youth on a product, watching
Those pale youthful days, running from an inland village
To a coastal factory, all the way to a store shelf in the U.S.A.
Fatigue and industrial diseases are accumulating in the lungs
Those hints: menstruation no longer comes on time
Violent coughing, she sees far away from the factory in the development zone
Green lychee trees being hacked down, the machines by her side
Are trembling . . . she rubs her red swollen eyes, positions herself
Between the moving products on the assembly line (*Poems Scattered* 25)

The interlaced references and images in the poem suggest intertwined connections between the female migrant worker's fatigue, sickness, and the products sold on the U.S. market, and between economic globalization and environmental destruction in China, especially in the countryside and the so-called economic development zones. As Alaimo eloquently argues, matters of environmental concern are always

> simultaneously local and global, personal and political, practical and philo-sophical. Although trans-corporeality as the transit between body and environ-ment is exceedingly local, tracing a toxic substance from production to con-sumption often reveals global networks of social injustice, lax regulations, and environmental degradation. (*Bodily Natures* 15)

By following the movement of migrant laborers from an inland village to a coastal factory, and the movement of the factory's products to the stores in the United States, and by juxtaposing the images of the coughing migrant worker in the factory with the greasy sobbing water and felled lychee trees in the development zone, Zheng depicts precisely the kind of "trans-corporeal-ity" that exposes global networks of social injustice and environmental deg-radation. Yet, the Global North remains safely distant from the immediate impact of such social injustice and environmental destruction.

Moreover, Zheng's poems such as "Lungs" highlight the fact that the impact of toxic environments on migrant workers ripples beyond individuals and production sites to the countryside. In fact, the migrant worker's "lungs" in the poem can be understood as "the proletarian lung within the networks of nature/culture, which are '*simultaneously real, like nature, narrated, like discourse, and collective, like society*'" (Latour qtd. in Alaimo *Bodily Natures* 28).[11] Effectively deploying the image of diseased "lungs" as an ex-tended metaphor in the poem, Zheng stretches the complex socio-economic ecological web even wider, and the impact of environmental injustice against migrant workers uncontainable to factories.

His slow, labored, heavy breathing, clogged lungs
Moving in his body are welding dust, lead dust, cement dust . . . firmly and tenaciously clutching
On the soft, frail leaves of lungs for life, like an iron nail inserted into an impoverished, lowly, insignificant body of flesh
His diseased lungs are wheezing violently in this industrial age, painful, agitated
Sounds of coughing spread among vulnerable bodies, shredding hope dim like the flicker on a cigarette butt
Theirs are lungs from the countryside, lungs from fields of poor crops, or one or two pairs of
Fateful lungs, diseased lungs, rotting lungs
Vocational diseases weigh down lower the countryside's low chimneys
Migrant workers' children are school dropouts, they stand still, lost in the smoke of small fires
The wet wood she puts into the stove resembles their father's lungs filled with dusts
Coughing convulsively

. . . .
I watch dust-clogged lungs in my life: one family diminishes in sunset
Their nearly dried-up life shakes with the sound of coughing heavy as lead
Blotchy like mountain slopes after tree-felling and mining
Agony and ugliness exposed. (*Pedestrian* 38)

While the "dust-clogged lungs" can be understood in terms of "the 'proletarian lung'," which Alaimo contends exposes the environmental and physiological aspects of class oppression, what they reveal in Zheng's poem is more than the fact of "how sociopolitical forces generate landscapes that infiltrate human bodies" (*Bodily Natures* 28). Zheng tactfully links the penetration of industrial dusts into the migrant worker's lungs to the disintegration of migrant workers' families and the wreckage of plunder of natural resources in the countryside. In doing so, she situates migrant workers' vocational diseases and the collapse of their families in the context of larger social and environmental problems, resulting from interrelated economic globalization, local economic development, and industrialization, which have led to the displacement of millions of migrant workers like the father dying of industrial disease in a factory far away from his family.

AGENCY AND A "TRANS-CORPOREAL" POETICS

As she further investigates the predicament of migrant workers, Zheng links their labor and work environment to globalization, and suggests that migrant labor in China reflects an unequal relationship between the mutually constituted Global North and Global South. In her poem "The Age of Industry," Zheng calls critical attention to the ethical, social, and environmental issues underlying this unequal yet interconnected relationship. She juxtaposes the transnational networks of industry and commerce with a variety of local dialects in a factory to highlight the connection between migrant labor in China and economic globalization, which has brought about a new era of industrialization in China and an emergent Global South within the country. The ironic tone of the speaker underlines the power of transnational capital of the Global North, whose transformative impact on people's lives and the environment in China is implied in the scene of a factory, which is contextualized in networks of the global market:

The Japanese machines in the American-invested factories carry iron produced
In Brazilian mines; lathe tools from Germany reshape the coastal lines
Of France; store shelves in South Korea are full of Italian labels
Belgium is waiting in the corner to be sold; Spain and Singapore
Are being inspected; Russia has been put into the warehouse by transporters; Africa
Is standing in the open-air field as natural resources; the orders from Chile are narrow and long
Like the shape of its territory; my Sichuan dialect is somewhat conservative, the

Xiangxi dialect
Is more difficult to understand; the Minnan dialect of Fujian is conversing with people
from Taiwan;
Cantonese-speaking Hong Kong is only a half-way station; if I'm willing
I could arrange India, Afghanistan, Pakistan in the vicinity
Of Australia, put Iraq and USA side by side
Move Israel to the middle of the Caribbean nations
Make England and Argentina shake hands, Japan and Mexico . . . .
In this industrial age, I am keeping busy every day
For the sake of peacefully arranging the world in a factory.
(*Poems Scattered* 57–58)

Embedded in the entwined operations of global economic networks is an
implied divide between the Global South and North, as Africa becomes
standing "natural resources," as tens of thousands of workers from various
parts of China gather in a factory whose products are shipped to countries
mostly in the Global North. The local-global connections in Zheng's poems
implicate the Global North in the displacement, exploitation, and dehumanization of migrant workers, as well as in the pollution of their working
environments, which are invisible to the distant consumers of the products
produced at the cost of lives and ecosystems in the Global South.

Situated in an emergent Global South in China, Zheng's articulation of
the migrant workers' bodies in her poems can be better understood not only
in terms of "trans-corporeality," but also by means of the critical analysis
embedded in the images. As Alaimo contends, by insisting on the insepara-
bility of human corporeality from the environment, trans-corporeality can
generate "rich, complex modes of analysis that travel through the entangled
territories of material and discursive, natural and cultural, biological and
textual" ("Trans-Corporeal Feminisms" 238). In fact, a similar mode of analysis is embedded in Zheng's poems, particularly those about female migrant
workers. In the poem "Wedded," Zheng employs the double meanings of
"wed"—to marry and to unite—to situate the female migrant worker's body
reduced to dehumanizing labor intertwined with capitalism and industrialization. As she writes: "screws are wedded to the operating machines," "blue-
prints are married to products," "the industrial zone begets / Factories,
noises, prosperity through weddings," "life has wedded a high-capacity machine / To me, it's tumbling, rumbling on my body" (*Poems Scattered* 33).
The meaning of marriage for migrant workers has been altered as their bodies are reduced to machines and exploited for cheap labor. Precisely because
human corporeality is in part socially and environmentally constituted, the
gendered and classed body of female migrant workers in Zheng's poems
enacts the agency of protest and social critique in complex ways.

For female migrant workers, trans-corporeality entails a form of sexual
exploitation made possible in part by the emergent environments—a new
"landscape," or rather a new geography, resulting from economic globaliza-

tion. Zheng's portrayals of the lives of Chinese female migrant workers reveal that subjugation and exploitation of the female body are prominent characteristics of the peculiar living environments generated by sociopolitical, cultural, and economic forces, including transnational capital. If the notion of the "proletarian lung" illustrates Alaimo's "conception of trans-corporeality, in that the human body is never a rigid enclosed, protected entity, but is vulnerable to the substances and flows of its environments, which may include industrial environments and their social/economic forces" (*Bodily Natures* 28), then the corporeality of female migrant workers as portrayed in Zheng's poems demonstrates that the harmful environments for those workers are not merely "industrial" toxins. In her recent volume *Female Migrant Workers* (2012), a collection of poems and prose, Zheng portrays the lives of nearly hundred women whose individuality is erased in the factories, where they are known by their respective work numbers. Zheng herself was called "No. 245" and sometimes referred to by the position of her seat on the assembly line. In this collection, she reclaims the humanity and individuality of female migrant workers by portraying them otherwise. She makes their lives known and their voice heard, and restores their individual names as titles of the poems about them. But even as she seeks to redeem the individual identity of each female migrant worker, the characteristics of each woman's life reflects a larger pattern of migrant workers' plight. This pattern reveals that the family structure of migrant workers is collapsing, and the social fabric of communities in their home villages is deteriorating as young people leave for the cities searching for a better life, often ending up as migrant workers, or in some cases becoming prostitutes, while remaining outsiders in the city, living on the margins of mainstream society.

At once corporeal and social, personal and collective, the female migrant worker's body, as depicted in Zheng's poems is reconstituted by its specific social, material, cultural, and constructed environments. Uprooted and living as singles far away from home, young female migrant workers are vulnerable to sexual exploitation as well as toxic materials in their working environments. Without the protection of families and communities, young female migrant workers often fall prey to male bosses and co-workers. Among the female migrant workers, some are minors, as shown in "The Girl from Guizhou." Although she is only "thirteen (or even younger)," "life has become merely a matter of survival" for her in "the factory of more than 10,000 workers," where "she is like a duckweed on the sea" (*Female Migrant Workers* 49). In the poem "The Child Migrant Worker from Cold Mountain," a fourteen-year-old girl experiences the same kind of unending fatigue on the assembly line as grown-up workers do. Moreover, these young women suffer another layer of gendered oppression and exploitation because of sexism and the lack of protective family and community networks. The young worker from Cold Mountain reveals that another girl "even though she is younger /

has to sleep with men at night" (*Female Migrant Workers* 52). In "Xiao Qing," the seventeen-year-old girl was "like a drop of dew on an ear of rice," or "the bright moon above the bamboo grove behind her house," "dropped in the city's endless complexes of buildings [. . .]" (*Female Migrant Workers* 53). Within a few years, Xiao Qing has lost her dreams as her body and spirit "have fallen into an abyss." Now living in a ghetto-like "village-in-the-city," her prostituted, violated, diseased body is "like the whole society / corrupt and rotting radiant and sumptuous / . . ." (*Female Migrant Workers* 54). These young women's lives reveal another multifaceted dimension of trans-corporeality—the "time-space" where female migrant workers' bodies are not only invaded by toxins in the environment, or subjugated as cheap labor, but also sexually exploited and violated. Their gendered exploitation suggests that as we seek to expand the ethical, political, and epistemological purview of environmental studies through the concept of "trans-corporeality," the gendered "proletarian" body must be a central concern in critical investigations in the ethics and politics of environmental justice. Like the migrant workers' "dust clogged lungs," the female migrant worker's body in Zheng's poems is *simultaneously real, like nature, narrated, like discourse, and collective, like society*," to once more quote Bruno Latour (*We Have Never Been Modern* 6). With such multiple, heterogeneous attributes and formations, the migrant worker's body enacts the agency of social critique and protest, an agency that is denied to the oppressed, exploited, and silenced subject.

In poems such as "Li Yan" and "Chen Fang" about married female migrant workers, Zheng highlights another profound social problem characteristic of the Global South in China, as implied in the workers' broken marriages and absent parenting for their children. In "Li Yan," the young woman lives away from her husband and child, while trying to have a family as a migrant worker. She arrived in Guangdong from Hubei in 1996 "with a young girl's bright dreams." In 1999, to "escape officers of family planning," she went into hiding in the bamboo forest in Sichuan, "eight-months pregnant her husband far away in Guangdong / Her parents far away in Hubei only the rain in Sichuan" keeps her company. In 2000, after taking "her seven-month son to Hubei," she returned to the factory in Guangdong (*Female* 44). In 2007, her estranged husband fell in love with another woman and divorced her. Li Yan left with a broken heart the place where she had been a migrant worker for ten years to start a new life again in another town. In 2008 she remarried someone who, like her, had been divorced. The birth of her daughter in 2009 seems to mark the beginning of a promising life in a new place, where she is "seeking a small quiet vessel for her migrant soul" (*Female* 45), yet she remains far away from her son left in Hubei. While Li Yan claims the center stage in this poem, her virtually "orphaned" son left with her aging parents in the countryside of another province, is more than a background reference. The deprivation of a family life is integral to the emergent manufactured and manufacturing landscape of enormous factory complexes in China, which are production sites for the world market.

The children with absentee migrant-worker parents appears in a number of Zheng's poems, and "migrant children" is a central concern in her other writings, including her essay "Elegy for the Village" reviewing Luo Ruiping's collection of poems entitled *Women's Village* (2010). In this essay, Zheng examines the connection between migrant workers and the "hollowing of villages" (n.p.). Li Yan's seven-month-old son left behind in Hubei evokes the drop-out children in the countryside of the sick migrant-worker father in "Lungs." These deserted children, like the deserted villages of migrant workers, are an integral part of the impact of economic globalization. They constitute part of a Global South in China produced by interrelated local and global economic and social forces. If "tracing a toxic substance from production to consumption often reveals global networks of social injustice, lax regulations, and environmental degradation," as Alaimo argues (*Bodily Natures* 15), then tracing social elements resulting from the local environments transformed by economic globalization, I would contend, can reveal broader effects of "global networks of social injustice" rendered invisible by marginalization of migrant workers' lives beyond the immediate production sites, and outside the purview of environmental justice discourses that focus mainly on toxins and their impact on the corporeal. The capaciousness of Alaimo's conception of trans-corporeality offers a complex approach to investigating matters of the environment, an approach that includes both the material and the socio-cultural as "actors" in the formation of an industrialized Global South in China.

Zheng's poems such as "Chen Fang" urge the reader to take into account the corporeal and the psychological, the material and the social, the environmental and the familial in shaping the lives of migrant workers through Chen Fang's situation. Although married with two children, Chen Fang still lives like a single woman, working on the assembly line where she started nine years ago:

> Hidden under her youthful face
> Is a seasoned heart
> Lethargic at the worn-out machine
> Young yet aging like her gentle sighs
> Quiet voice low bending head
> Foot steps fatigued and empty she is from Hubei
> Mother of two children her twenty-six-year-
> Old husband is a transporter in a molding factory
> In another city he left her a year ago
> Eight years before that she and he crammed into a van
> Arrived here from a village in Hubei four years ago
> They returned home to be married a fifteen-day holiday

Pregnant then giving birth she returned home for a year

. . . .

—it has been nine years she is used to

The 15-day shift, installing 20,000 slingshots a day

. . . .

Used to fatigue and drowsiness used to the breeze from the sea
Used to going without breakfast this girl Chen Fang
Arriving here in 1998 in 2007 she still remains
On the assembly line she feels a faint pain
In the stomach like the pain the felled trees feel
Pressing a hand on her ulcerous stomach she went home
Saying she'd return when she got well
She talked about the changes around here describing
The cold river turning smelly remembering
The sound its running water used to make the aching stomach
Makes her somewhat apprehensive and a little sad. (*Female* 186)

While Chen Fang does not complain, her quiet acceptance of her situation—
fatigue, illness, separation from her children, and disintegration of her family—actually exposes larger social and environmental problems beyond labor
exploitation. Zheng arranges the lines of the poem in such a way so that her
long hours of tedious work on the assembly line is linked to the pain Chen
Fang feels in her stomach, and this pain in her body is connected to "the pain
the felled trees feel." At the same time, the deteriorating health of Chen Fang
is juxtaposed with environmental degradation—the disappearing lychee
trees, the once running river now having turned smelly and stagnant. The
"time-space" of the trans-corporeality of Chen Fang's body and its working
and living environments are further extended to other locations, including the
molding factory where Chen's first husband works as a transporter, and the
village in Hubei, where their two children have been left. Moreover, like the
subjugated, dehumanized body of the female migrant worker, which is metaphorically turned into an iron nail on the wall, and similar to the migrant-
worker father's dust-clogged lungs and to Xiao Qing's violated, infected,
infectious, and ailing body, Chen Fang's exploited, fatigued, and diseased
body takes on a form of agency of social exposure and critique.

In the prose piece "Handwritten Notes 11: Unnatural Deaths of Female
Migrant Workers," collected in *Female Migrant Workers*, Zheng calls public
attention to migrant workers' wounded bodies and the high rate of "unnatural
death" among them. She herself was injured while working in a hardware
factory when a machine tore the nail from her index finger. Since then she is
terrified of machines and the excruciating pain remains vivid. When she read
a newspaper report saying that the annual number of injuries of broken
fingers in the Pearl River Delta region in Guangdong alone is more than
40,000, she felt the inadequacy yet necessary significance of words for making known the hazardous working conditions, especially the high rate of

"unnatural death" among migrant workers. She notes that between the years 1996 and 2006, nine migrant workers from the impoverished River Mouth Village died unnaturally. 80 percent of the village's relatively small population of 1,300 are migrant workers, who were young and healthy when they left. Hence River Mouth Village is also known as "Migrant Workers' Village," one of the countless villages of migrant workers in China (*Female* 179). It would be impossible to know the actual number of unnatural deaths between 1996 and 2006 because they were not reported. When more than ten Foxconn suicides happened within one year, they attracted a lot of media attention, but only for a while. Zheng states, "I know suicide incidents continued at a Foxconn factory after that. But those workers disappeared from the news into oblivion in this country. Such suicides in this age do not make 'news' anymore. The dead are eventually digested completely in the stomach of our era" (*Female* 180). All she could do, Zheng states, is "write poems to reclaim their lives, which are a musical note of our times, though for some people it is not in syncopation with our era" (*Female* 180). Dedicated to migrant workers' lives and the Global South in China, which includes migrant workers' villages, Zheng's poems refuse to celebrate economic development as "progress," but rather, urge the reader to consider at whose expense and at what social and environmental cost is "prosperity" obtained and products made for the markets of the Global North.

TRANSMUTATIONS OF THE "PROLETARIAN" BODY

Zheng's poems about migrant workers accomplish more than bearing witness to social and environmental injustice, and more than reclaiming disappeared, marginalized lives at the bottom of society. Even though Zheng often compares the migrant worker to "iron" that is mute and subject to being cut, slated, melted, hammered, reshaped, and used, she observes the transmutation of matter for inspiration about the transformation of migrant workers' bodies and spirits in order to resist and overcome oppression. Thus, a transformative "trans-corporeality" often takes place in her poems. In her prose poem "Iron," Zheng writes: "Life of migrant labor is like an acid rain, it erodes our flesh, soul, ideals, dreams, but not a liquid heart, which is stronger than iron and steel" (*Pedestrian* 177). The indomitable "liquid heart" is a unique "trans-corporeal" image and concept emerging from Zheng's observation of iron melting in the furnace while working in a hardware factory:

> I often observe for a long time as a piece of iron changes in the furnace fire . . . . As I watch, the scorching heat turns red, translucent red, clear like a teardrop, making my eyes water. When my tears fall on the searing iron, they disappear immediately. Even now I insist on thinking that my tears were not evaporated by the high heat from the furnace fire, but rather disappeared into

the iron, becoming part of it. Tears are the hardest matter in the world; they
have a soft yet indestructible power. (*Pedestrian* 176)

In the "contact zone" where human corporeality and matter meet, a transmutation of both takes place in their respective, interrelated becomings. This
trans-corporeality in Zheng's poem mobilizes a "corporeal agency" (Alaimo's phrases in "Trans-Corporeality" 250) when the worker's body and
spirit become one with the light of the melting iron in the furnace, the "liquid
fire" (*Pedestrian* 176). Zheng articulates a transformative corporeal agency
in another poem entitled "Written in Sickness" through the speaker, who
states that the migrant workers' "muscle" that "props up half of the country
on the verge of collapsing," will "rebel" (*Selected Poems* 27).

Theories of "material agency," or "agency without subjects" developed
by material and ecofeminist critics such as Karen Barad and Alaimo open up
new possibilities for investigating various forms of agency as embedded in
Zheng's poems. In her essay "Posthumanist Performativity: Toward an
Understanding of How Matter Comes to Matter," Barad argues that "matter
does not refer to a fixed substance; rather, *matter is substance in its intra-
active becoming—not a thing, but a doing, a congealing of agency. Matter is
a stabilizing and destabilizing process of iterative intra-activity*" (139). She
adds, "The dynamics of intra-activity entails matter as an *active* 'agent' in its
ongoing materialization" (140). Agency, then, is not enacted solely by intentional subjects, but rather, mobilized by the intra-activity of matter in the
process of its becoming. In a similar vein, Alaimo in her essay "Trans-
Corporeal Feminisms" contends that:

> Crucial ethical and political possibilities emerge from this literal 'contact
> zone' between human corporeality and more-than-human nature . . . . But by
> underscoring that 'trans' indicates movement across different sites, trans-cor-
> poreality opens up an epistemological 'space' that acknowledges the often
> unpredictable and unwanted actions of human bodies, non-human creatures,
> ecological systems, chemical agents, and other actors. (238)

Both Barad's and Alaimo's emphases on material agency without subjects
can be applied to examining the "often unpredictable and unwanted" effects
of the "destabilizing" corporeal agency of female migrant workers as portrayed by Zheng.

Trans-corporeality in Zheng's poem "They Who Demand Their Wages on
Their Knees" mobilizes the mute female migrant workers' bodies into an
unexpected rebellious action of protest and resistance:

> They flash by like ghosts at train stations
> At machines in the industrial zones in dirty rentals
> Their thin figures like blazes like white paper
> Like hair like air they have cut with their fingers

Iron films plastic . . . they are tired and numb
Looking like ghosts they are put into machines
Overalls assembly lines they are bright-eyed
Youthful they flash into the floods made up of themselves
The dark waves I can no longer tell them apart
Just as I'm indistinguishable standing among them only skin
Limbs movements blurred features one after another
Those innocent faces are ceaselessly grouped lined up
Forming ant colonies bee hives of toy factories they are
Smiling standing running bending curling up then each
Reduced to a pair of hands thighs
Becoming tightly tightened screws cut into pieces of iron
Pressed into plastic twisted into aluminum threads tailored to cloth
. . . .
They are from small towns villages hamlets they are smart
Clumsy they are timid weak . . .
Now they're kneeling on the ground facing huge bright glass doors and
windows
Black-uniformed security guards shiny cars New Year's greeneries
The golden factory signboard is glittering under the sunlight
They are on their knees at the factory gate holding a piece of cardboard
With clumsy handwritten words "Give Me My Blood & Sweat Pay"
They four are fearless kneeling at the factory gate
. . . . (*Female* 107–08)

Even though these female migrant workers are dragged away by security guards, their performance of "begging" in public is a powerful gesture of defiance, a rebellious demand for fair treatment, a protest against injustice. In contrast to their diminished status, being reduced to body parts—"skin," "a pair of hands," "thighs," and objects—"screws," "iron," and "plastic"—and disciplined into a uniform collective like "ant colonies" and "bee hives," the corporeal agency of the four courageous female workers is particularly unsettling. The trans-corporeal movement between the workers' bodies and the objects in the "contact zone" of the toy factory enact transformations of the subjugated and abused bodies. Female migrant workers who are becoming iron, plastic, tightened screws, and a uniform collective unexpectedly use their bodies to perform resistance that unsettles the apparently well-organized and disciplined docile image, or rather behavior, of migrant workers.

An agency of resistance mobilized by a trans-corporeal movement between the migrant worker's body and manufacturing tools and materials is more explicitly articulated in Zheng's poem "Disassembled." Zheng enacts the transformative agency for re-imagining and re-articulating the migrant worker's body, identity, and subjectivity, mobilizing a trans-corporeal movement as a process of becoming otherwise through the speaker as a migrant worker: "Disassemble."

I disassemble my bones, soul, blood, flesh, and heartbeat

Into screws, film, plastic parts, splinters, and hooks
Which are assembled, recombined, assigned standard labels, so that childhood
Dissolves into apparitions of memories, times past, and emotions. I disassemble
Love into blue prints, products, wages, overtime, I-owe-you pays, and insomnia
So, too, is the vertical society disassembled into horizontal disasters, villages,
countryside sorrows
If the flames in the furnace cannot ignite a piece of rusted iron . . . yet I
Am searching for meanings of life in a life filled with rust, those
Ideals and passions of the past are dismantled by the crushing power of the hardware
factory, which
Dissects a person into parts, screwed in a corner of society
How have certain diseases of industry penetrated our bodies
This misfortune belongs to our era, or the masses
. . . .
In this era I will disassemble myself into springs
Switch valves, wires, a steel needle, a street lamp
I'll once more return to the furnace, to forge myself
Into shape, to hack myself into a sharp nail
To be nailed on the wall of our era. (*Poems Scattered* 59)

A dehumanizing alteration of the self-disassembled by the hardware factory at the beginning of the poem undergoes a transmutation by the end. Identifying her corporeal self with pliable objects, the speaker re-imagines and re-invents herself into "a sharp nail," which is fundamentally different from the nails as tools in the factory, or the nail fixed on the wall like a hanger for the "master," as depicted in "An Iron Nail." Just as the self-determined trans-corporeal becoming in "Disassembled" and Zheng's other poems characterizes a poetics rooted in the migrant worker's experience, the speaker's determination to be "nailed on the wall of our era" suggests a keen sense of the contemporary epoch-changing historical moment. It also asserts a socially engaged ethical and aesthetic commitment of the poet and her poetry.

Embedded in the local-global connections in Zheng's poems about migrant workers and an emergent Global South in China are urgent environmental justice issues raised in a recent critical anthology, *American Studies, Ecocriticism, and Citizenship: Thinking and Acting in the Local and Global Commons* (2013), edited by Joni Adamson and Kimberly Ruffin. In their introduction, Adamson and Ruffin argue for "the urgency of coming into a 'middle place' or creating a 'methodological commons' where academic and public discourse about citizenship and belonging in both local and global contexts might become more accessible and clear, and thus, more transformative" (16). They seek to expand "notions of what constitutes 'the community of rights' and the 'rights of community' and how we might better support individuals and groups who are part of nations *and* planetary citizens in creating and enacting policies, laws, and community practices that will have positive ecological consequences around the globe" (16). Zheng Xiaoqiong's poems challenge us to expand notions of community and rights for social, ecological, environmental justice beyond national borders. The migrant workers and the Global South China Zheng represents urge us to confront the ways in which the predicaments of the Global South are in part produced by various forces of the Global North. Recognizing this intricate connection can help us better understand both the challenges in and urgency for "thinking and acting in the local and global commons."

NOTES

1. See James Pomfret. "Foxconn Worker Plunges to Death at China Plant: Report." *Reuters*. November 5, 2010. Web.; Malcolm Moore, (2010-05-16). "What Has Triggered the Suicide Cluster at Foxconn?" *Telegraph Blogs*. May 16, 2010. London: Blogs.telegraph.co.uk.; and Fiona Fam. "Foxconn Factories Are Labour Camps: Report." *South China Morning Post*. October 11, 2010. Web. Mimi Lau, "Struggle for Foxconn Girl Who Wanted to Die." *South China Morning Post* (Wuhan, Hubei, China). December 15, 2010. Web.

2. See for example, Jason Dean, "Apple, H-P to Examine Asian Supplier after String of Deaths at Factory." *The Wall Street Journal*. May 27, 2010. Web.

3. Tim Culpan, "Foxconn to Hire 400,000 China Workers within a Year," *Bloomberg*, 18 August 2010. http://www.businessweek.com/news/2010-08-18/foxconn-to-hire-400-000-china-workers-within-a-year.html. Web.

4. Thomas Dinges, "Foxconn Rides Partnership with Apple to Take 50 Percent of EMS Market in 2011," iSuppli, 27 July 2010. http://www.isuppli.com/Teardowns-Manufacturing-and-Pricing/News/Pages/Foxconn-Rides-Partnership-with-Apple-to-Take-50-Percent-of-EMS-Market-in-2011.aspx.

5. See Malcolm Moore, "'Mass Suicide' Protest at Apple Manufacturer Foxconn Factory." *The Telegraph* January 11, 2012. (*The Telegraph*, 30 July 2012. Web.)

6. Recently some of Zheng's poems have appeared in English translations. Five poems in English translated by Jonathan Stalling are published in *Chinese Literature Today* and three also translated by Stalling are included in *New Cathay: Contemporary Chinese Poetry, 1990–2012* edited by Ming Di (Tupelo Press, 2013).

7. See Zheng Xiaoqiong, Afterword, *Female Migrant Workers*, 255.

8. Zheng also mentions in the interview that "According to the Labor Law, 'all factories and companies are required to restrict overtime for workers, normally no more than one hour a day. Overtime can be extended to no more than three hours a day and no more than 36 hours a month under extraordinary circumstances and on condition that workers' health is not jeopardized.' In reality, few factories obey this law. The overtime in most factories is more than 80 hours, even over 100 hours in some factories. This has almost become a common phenomenon." See Zheng Xiaoqiong, Interview, "Regarding Peasant Workers and Foxconn." June 2, 2010. http://blog.sina.com.cn/s/blog_45a57d300100kmt2.html. Web.

9. All the English translations of Zheng Xiaoqiong's poems and words in this essay are by the author.

10. The sense of isolation and marginalization in the cities as expressed in the poem is partly the result of the residential status of peasants in the city, and partly the low social status of migrant workers. According to residential policies in China, those registered as countryside residents are not allowed to become regular city residents even though they have worked and lived in the cities for years. The policies in part are aimed at controlling urban overcrowding.

11. The quote within this citation of Stacy Alaimo is from Bruno Latour, *We Have Never Been Modern* (6).

WORKS CITED

Adamson, Joni and Kimberly Ruffin, Introduction. *American Studies, Ecocriticism, and Citizenship: Thinking and Acting in the Local and Global Commons*. New York: Routledge, 2013. 1–17. Print.

Alaimo, Stacy. *Bodily Natures: Science, Environment, and the Material Self*. Bloomington: Indiana UP, 2010. Print.

———. "Trans-Corporeal Feminisms and the Ethnical Space of Nature." In Alaimo and Hekman. 237–64. Print.

Alaimo, Stacy, and Susan Hekman, eds. *Material Feminisms*. Bloomington: Indiana UP, 2008. Print.

Barad, Karen. "Posthumanist Performativity: Toward an Understanding of How Matter Comes to Matter." In Alaimo and Hekman 120–54. Print.

Barboza, David. "IPhone Maker in China Is Under Fire After a Suicide." *New York Times*. 16 July 2009. Web.

Gong, Haomin. "Toward a New Leftist Ecocriticism in Postsocialist China: Reading the 'Poetry of Migrant Workers' as Ecopoetry." *China and New Left Visions: Political and Cultural Interventions*. Ed. Ban Wang and Jie Lu. Lanham, MD: Lexington, 2012. 139–57. Print.

Latour, Bruno. *We Have Never Been Modern*. Trans. Catherine Porter. Cambridge, MA: Harvard UP, 1993. Print.

Lau, Mimi. "Struggle for Foxconn Girl Who Wanted to Die." *South China Morning Post*. July 19, 2012. Web.

Moore, Malcolm. "'Mass Suicide' Protest at Apple Manufacturer Foxconn Factory." *The Telegraph*. July 30, 2012. Web.

Stalling, Jonathan. "Five Poems by Zheng Xiaoqiong." (Chinese and English Translations). *Chinese Literature Today* 1.1 (Summer 2010): 56–59. Print.

Students & Scholars Against Corporate Misbehaviour (SACM). "Workers as Machines: Military Management in Foxconn." 12 October 2010. Web.

Yang, Jincai. "Ecological Awareness in Contemporary Chinese Literature." *Neohelicon* 39 (2012): 107–18. Print.

Zhang, Qinghua. "Who Touches the Iron of the Age: On Zheng Xiaoqiong's Poetry." *Chinese Literature Today*. 1.2 (Summer 2010): 31–35. Print.

Zheng, Xiaoqiong. *Aches and Pains*. Literature of the Masses Publishing House, 2009. Print.

———. *Dark Night*. Beijing: Literature of the Masses Publishing House, 2008. Print.

———. "Elegy of the Village." August 4, 2010. Web. http://blog.sina.com.cn/s/blog_45a57d300100m5ky.html.

———. *Huang Maling Village*. Beijing: Long March Publishing House, 2007. Print.

———. Interview. "Regarding Peasant Workers and Foxconn." June 2, 2010. Web. http://blog.sina.com.cn/s/blog_45a57d300100kmt2.html.

———. *Female Migrant Workers*. Guangzhou: Flower City Publishing House, 2012. Print.

———. *Pedestrian Overpass*. Taibei, Taiwan: Tanshang Publications, 2009. Print.

———. *Poems Scattered on the Machine*. Beijing: China Society Publishing House, 2009. Print.

———. *Pure Plants*. Guangzhou: Flower City Publishing House, 2011. Print.

———. *Selected Poems by Zheng Xiaoqiong*. Guangzhou: Flower City Publishing House, 2008. Print.

———. *The Depths of Night*. Beijing: Literature of the Masses Publishing House, 2006. Print.

———. *Two Villages*. Beijing: Literature of the Masses Publishing House, 2007. Print.

Acknowledgment: I am grateful to Zheng Xiaoqiong and her publishers for granting me the permission to translate and publish the translations of her poems included in this essay.

Chapter Four

Literary Isomorphism and the Malayan and Caribbean Archipelagos

Christopher Lloyd De Shield

Édouard Glissant opens *Caribbean Discourse* with a straightforward declaration of specificity that seems to forestall any attempt at archipelagic comparison: "Martinique is not a Polynesian Island" (1). With this provocative opening, Glissant rebuts past and present colonists and tourists who apprehend the land in a particularly insensible fashion: namely, as existing solely for the pleasure of the privileged. Martinique, he seems to insist, is not simply one of "many islands"—interchangeable and dismissible as a group. As such, Glissant's qualification undermines colonial and touristic discourses of exploitation.

However, it is only on a first reading that Glissant's invocation of Polynesia insists on Caribbean (more specifically Martiniquan) difference. A second reading reveals how Glissant simultaneously invokes the shared predicament of (post)colonial islands apprehended within colonial discourse. In articulating the "opacity" of the Caribbean island in question, he seems to reject the invitation for comparison by preventing the two regions' inclusion within the same intellectual frame; but at the same time—in rejecting this colonial comprehension of tropical islands—Glissant necessarily implies a counter-discursive equivalence. That is, insisting on Polynesian and Caribbean islands' shared rejection of colonial comparison ironically reifies a comparative project; the archipelagos begin to resemble one another in the similarity of their strategic claims to postcolonial opacity. Thus, Glissant's insistence on the specificity of the island immediately plunges it into relation with other islands.

This relationship between specificity and relationality, or comparison, is almost a disavowed dialectic within postcolonial studies; only recently have attempts to theorize the field from regionalist perspectives been re-examined.[1] Comparative approaches within postcolonial studies are frequently greeted with suspicion, risking charges of a- and de-historicization, or de-contextualization.[2]

Critics of the comparatist approach argue that the comparative gesture ironically invokes the specter of colonial history (Robinson, 2011). Jennifer Robinson argues that such a methodological maneuver presents a "significant dilemma"; that is, "comparativism—the ability to cast one's eye across a range of disparate cases and consider them equivalent for analytical purposes—is itself a profoundly colonial inheritance" (126). The comparative methodology mirrors colonial logic in its desire and power to apprehend diverse regions from a domain of intellectual and material privilege.

What have been less analyzed within postcolonial studies, however, are the perils of *over*-contextualization. In *Absolutely Postcolonial*, Peter Hallward clearly spells out two real problems with the fetishization of particularity in postcolonial studies: he declares how "it should be obvious that mere insistence on particularity (on the this-ness of things) cannot resolve any *theoretical* question whatsoever" (39). Moreover, he reveals the link between nearly unqualified affirmations of locality and specificity and, "exuberantly nativist essentialism" (36–37). Hallward argues that there is only a small step to jump between celebrations of cultural specificity—what he calls "bland contextual respect"—and literary judgment based on cultural *authenticity*, a form of literary assessment that postcolonial criticism incessantly disavows (37).[3]

If—as Peter Hallward has argued—Glissant's work is, in fact, broadly opposed to the current postcolonial consensus advocating increasingly greater specificity in postcolonial scholarship, then reading his opening invocation as a double-gesture is an appropriate one. Such a double-gesture highlights a useful corrective to the dominant mode of postcolonial scholarship overly preoccupied with center-periphery interactions to the exclusion of South-South, or periphery-periphery, global flows.

In his essay *Beyond the Straits: Postcolonial Allegories of the Globe*, Peter Hulme offers a critique of comparativism that reveals its colonial genealogy (47). Hulme writes, "[o]n one level, globality—even in a restricted sense of the term—is clearly directed at the attainment of military and commercial power"[4]; therefore he opens his essay questioning the change to a larger scale given "postcolonial studies' fundamental commitment to ideas of the local and the marginal" (47). Detouring from Hulme, I ask here how one might invoke a Glissantian double-gesture, approaching specific, national or local contexts *and* broad, comparative, transnational flows within the same dialectic.

To this end, the concept of *analogous structures*—a technical term borrowed from the biological sciences—is apposite.[5] Extrapolating the definition from the biological to the humanities and social sciences, it refers to: any comparison of *structures* with different *evolutionary pathways* but having the same essential *function*.[6] Similarly, analyzing the functional element within postcolonial literary studies identifies analogies on the level of performance between literary works appearing out of very different socio-cultural environments.

But is it possible to give an account for such similarities, without relying on a standard "postcolonial" explanation—that similarities in literary works of postcolonial regions simply result from resistance to colonial hegemony, and the chief task of the critic is to locate instances of such resistance? In this, albeit crude and grossly caricatured perspective of postcolonial practice—that nonetheless persists in certain quarters—the true merit of a work of literature might be measured by tallying up instances of the "resistance" they display; or, as in Peter Hallward's caution (and Réda Bensmaïa's complaint),[7] the danger lies in taking perceived authenticity of representation as an example of resistance simply for the fact of presenting a non-European subjectivity from the margins.

Perhaps a change in methodological perspective can cure this predilection. To illustrate this claim, I will track the figuring of Southeast Asia (including the "Malay Archipelago") and the Caribbean (both continental and island) as analogous structures within three different, but interlinked, intellectual paradigms: the colonial imaginary, the postcolonial literary, and postcolonial ecocriticism.

THE COLONIAL IMAGINARY

Despite the lack of contemporary critical attention (resulting from methodological biases), both geographical regions have been historically and imaginatively comprehended (etymologically, "grasped together") in the colonial imaginary.[8] Recall how, historically, the Caribbean and Southeast Asia are analogous in colonial European thought because—while they developed independently—they served the same function. That is, within the logic of colonial mercantilism, *both regions* were perceived and conceived of as regions of *fecundity* and *commodity*, providing gold and spice, sugar, and tobacco, the commodities of empire and power in colonial times. Analyses of either region, at least since Said's seminal *Orientalism* (1978), have pointed to the regions' providing an "other" against which Europeans defined themselves. And Richard Grove's influential *Green Imperialism* (1995) has clearly demonstrated how the tropical islands of both regions provided a stimulating encounter that led to the development of the natural sciences and Western thought.

But one of the most obvious analogies between the two regions—within the paradigm of the colonial imaginary—is that both regions, the Caribbean and Southeast Asia, have had the same name: "Indies." The significance of this name for the regions is typically under-theorized. "Indies" is too easily dismissed as a kind of geographical malapropism on the part of early European explorers that has evolved into a misnomer that stuck. The definition and etymology of the word "Indies" yields, from the Oxford English Dictionary (OED), two definitions, one obsolete. The first is the standard usage, that "Indies" is a "plural adaptation of 'India' and signifies India and the adjacent regions and islands, and also those lands of the Western Hemisphere discovered by Europeans in the fifteenth and sixteenth centuries, originally supposed to be part of the former"; the OED records how "with the progress of geographical knowledge the two were distinguished as East Indies and West Indies" (OED).

Now, in addition to this standard definition, the OED notes a now obsolete usage of the word: that is, it is "used allusively for a region or place yielding great wealth or to which profitable voyages may be made." This latter—now obsolete—usage of the word is much more interesting. It is one that would have had currency during the age of European exploration. Naturally, the OED also documents early recorded instances in the English language of the word with this usage; the earliest listed in the OED happens to be found in Shakespeare, specifically his *Merry Wives of Windsor* (1598). In the scene in which it occurs, we have Falstaff discussing two women whom he describes as his personal "Indies":

> Fal.: Heree's a Letter to her. Heeres another to misteris Page. Who euen now gaue me good eies too, examined my exteriors with such a greedy intention, with the beames of her beautie, that it seemed as she would a scorged me vp like a burning glasse. Here is another Letter to her, shee beares the purse too. They shall be Excheckers to me, and Ile be cheaters to them both. They shall be my East and West Indies, and Ile trade to them both. Heere beare thou this Letter to mistresse Foord. And thou this to mistresse Page. Weelethriue Lads, we will thriue. (I.iii.60-8)

Here, we get a gendered construction in which the riches Falstaff can take from both women is analogous to the great wealth that may be "cheated" from both Indies.[9] As the OED indicates, an "Indies" for the Elizabethan colonist is simply a source of great wealth; the geographical location of the Indies is of no particular importance so long as it is *apprehended* as such.

Strangely not listed in the OED—given that it must be earlier—is another instance of this use of the Indies in Sir Walter Raleigh's *Account of the Discovery of Guiana* of 1595:

[. . .] I have anatomized the rest of the sea towns as well of Nicaragua, Yucatan, Nueva España, and the islands, as those of the inland, and by what means they may be best invaded, as far as any mean judgment may comprehend. But I hope it shall appear that there is a way found to answer every man's longing; a better Indies for her Majesty than the king of Spain hath any [. . .] (9)

Raleigh's account is of a journey performed in 1595 that undoubtedly provided source-material for Shakespeare's play. His usage renders the term explicit: "Indies" is no specific geographical location, it is a project and an ideological position; installing an extractive colonial machine, one can *make* an Indies.

In John Donne's "The Sun Rising" from *Songs and Sonnets* (1601) we get another literary application in which the persona invokes the metaphor of riches to describe his lover. In the poem, the persona addresses the rising sun that intrudes upon the pair of lovers lying in bed, with a commanding, "Look, and tomorrow late, tell me / Whether both the Indias of spice and mine / Be where thou leftest them, or lie here with me" (lines 16–18). Donne's persona provides a metaphor for the riches he enjoys in his lover's arms in the riches of gold and spice got from the colonies—East and West. The object of his description is his lover, but if we reverse the metaphor, we see that Donne has likened the exploitation of a geographical region to seduction and intercourse—conventional tropes of the colonial penetration of new lands at least since Raleigh declared "Guiana is a contrey that hath yet her maydenhead" (210).

These two regions are linked in the Colonial European imaginary through a violent *comprehension* characterized by that remarkable tendency to convert difference into the same. Just as Joseph Roach insists we consider how the "New World was not discovered in the Caribbean, but one was truly invented there" (4), it is worth exploring how certain European powers attempted to create what they wanted to find (by first erasing what existed). In *Routes and Roots*, Elizabeth DeLoughrey points to the scope of the imaginary in such comprehensions when she references Marco Polo's narrative, which, because it "had already described great archipelagos in Asia," made Columbus's Caribbean landfall seem predestined, so that what occurs is "a collapse of time-space between Antillean and Asian islands" (11).

We can trace this colonial *comprehension* in cartographic representations and in the transportation and transplantation of flora and fauna between regions: animals and fruits, plants and people.

THE POSTCOLONIAL LITERARY

Within postcolonial literary studies we see, ironically, certain inheritances from the colonial imaginary. While the two regions might not be perceived as analogous in native imaginaries prior to the struggle for independence, when approached via postcolonialism, both regions are *in practice* comprehended together under a single hegemonic category, ironically replicating the colonial tendency.

Rey Chow takes the Eurocentrism inherent to the comparative method to task in her essay "The Old/New Question of Comparison in Literary Studies: A Post-European Perspective" (2004). Chow advances the notion that, in comparative literature, the formulation "Europe and its Others" is fully in play, where national literatures are aligned and inserted into a European grid-of-reference "in subsequent and subordinate fashion" (294). In *The Postcolonial Exotic*, Graham Huggan notes how, tellingly, institutionalized postcolonialism operates under domination by the former colonial metropoles (4). And both Huggan, and Chow, have shown, in their own ways, how postcolonialism can induce a tendency so that texts are pre-read for the "resistance" they show. In postcolonialism, aggrandized claims to marginality or minority effect a strategic accumulation of "lack" (for Chow) or "exoticization" (for Huggan), valuable for the academic market.

As in the second reading of Glissant opening this discussion, at its most universal and reductive, postcolonialism apprehends the Caribbean and the Malayan Archipelagos as equivalent signs of difference or resistance for the West. In so doing, postcolonialism operates at the totalitarian level; it renders whole communities flat. The Caribbean and Southeast Asia then become new products, intellectual commodities for increasing knowledge and consequently, the power of, especially, the West.[10] Glissant points out a similar danger in his *Poetics of Relation* regarding orientalizing poets: "the elsewhere, full of diversity . . . somehow always ends up contributing to the glorification of a sovereign here" (37).

While it would seem logical to turn to the nation for shelter from the globalizing gaze of theory, many have recognized problems in the privileging of the novel and the nation as analytical categories in postcolonial studies. In this postcolonial literary framework, each country's literary output is set side by side with the only apparent justification being a shared project of decolonization, thereby re-invoking the suggestion of strictly analogous situations. In Pablo Mukherjee's phraseology—in *Postcolonial Environments*—we risk "wip[ing] out all differences between [regions] and see Moscow and Mogadishu, or Mumbai and Kinshasa as interchangeable entities" (7). This is exactly why some theorists of the Caribbean and Southeast Asia consider postcoloniality a hubristic attempt at framing the complexity of the regions (Lim 2). Like Aijaz Ahmad—in his famous critique of Frederic Jameson—

they scorn the tendency to "submerge" the "enormous cultural heterogeneity of social formations" into a "singular identity" (10).

Of course, much has been said about the perceived inadequacy of the term "postcolonial," but what is worth reiterating here is how these regions remain effectively analogous moving from the "universal" colonial European perspective through to the postcolonial literary.

CONTINENTAL CARIBBEAN: PENINSULAR SOUTHEAST ASIA

Consider the critical reception of *Beka Lamb,* an independence-era text by the Belizean author Zee Edgell. While *Beka Lamb* aims to "straightforwardly put on record a time and a place and a group of people: the life of a creole family in Belize City at the end of the Second World War" (Deveson 1187), Edgell's attempts are still evaluated within academe according to theoretical perspectives that aggrandize specific signs of marginalization while ignoring the local historical information that ironizes or even contradicts these conclusions.[11]

At the wake of the protagonist's grandmother a conflict erupts in the family in which the mother, who wants the family to progress from superstition, accuses Beka of associating too much with the "Caribs" (or, "Garífuna"—an indigenous ethnic group settling the coastal Caribbean of African and Caribbean descent). In the midst of this quarrel Beka interjects, quoting her Garífuna teacher, shouting: "When I grow up I am going to marry a Carib!" (68). Beka Lamb's mother is outraged and slaps her across the face. Literary critic Richard Patteson argues that this scene precipitates the climax of the novel. He says,

> The Caribs, though they are mentioned only a few times in *Beka Lamb,* constitute the novel's symbolic heart. The Caribs through their name and their ancestry are an indivisible link between Belize and the Caribbean past, including its pre-colonial history. One of the most marginalized peoples on earth, the black Caribs trace their ancestry back to Amerindians. (63)

Patteson continues stating that "the story of the Caribs is the story of Belize in microcosm" and that the Caribs "are in a sense the truest creoles" (64).

It is important to recognize this slippage between indigenous marginality, Creole subjectivity, and the unitary nation. Patteson's strategic conflation of Carib marginality with Creole subjectivity endorses the critical tendency to ascribe power to lack (by first engaging it within the realm of first-world theory). Marginalization has a strategic value which the literary scholar invokes to criticize colonialism. But without taking into consideration the text's cultural specificity and examining the concrete historical context in which the novel is situated, this is not an unproblematic invocation. Instead,

Patteson invokes an "other" described purely in terms of its indigenous or Africanized alterity. Indeed, concluding the paragraph, Patteson says simply, "both those peoples [Amerindians and Africans] had been exploited, manipulated, and brutalized by the hegemonic European powers" (64) and he can add nothing about Garífuna cultural relationship with the land, inside the country or with other ethnic groups.

However, literary scholars are not alone in exploiting the power of marginalization. Postcolonial writers and nations are frequently complicit in suggesting readings that appropriate "the indigenous" for specific ideological ends. In 1957, the year of Malaysia's independence, government decided to re-publish a book first written in 1937 by Ishak Haji Muhammad, titled *Putera Gunung Tahan* (The Prince of Mount Tahan). In the year it was published (1937), the nation's rulers were dealing with the campaign for a National Park and the novel is written as a response to territorial concessions that were seen by some as the "ultimate surrender by the Malays of their sacred rights" (Kathiramby-Wells, 246). *The Prince of Mount Tahan* is a thinly disguised satirical allegory embraced by the new nation. The novel features an epistemological contest between a Malay mountain fairy prince and a British colonial scientist, in which the British scientist is on foreign ground—his efforts to usurp the prince of his land result only in his self-conflagration.[12] The Malay Prince is a literal "bumiputra" (prince of the soil) which establishes a "natural" connection to the soil (indigenous to the point of being magically inexplicable).

The magical sovereignty of the Malay Prince is such that nature reveals its secrets to him, and aids him in his quest to confound British attempts at commercializing a natural monument (a hill-top colonial rest-station was proposed on the Mountain on which this fictional prince resides). The narrator directly addresses readers, informing them of the differences between rural dwellers, or jungle peoples, and city folk: "We are more noble than they are, even though they are supposed to be more advanced" (19–20).

As Kathiramby-Wells describes it, "the cultural values the forest represented had renewed meaning for Malay national identity" and the novel is "a poignant expression of the Malay reaction to external intrusions on their physical space, defined by the forest" (247). At the same time however, the discourses at play include that of re-hierarchizing society, assimilating the *Orang Asli* (literally 'original people') or the indigenous, into the sociocultural and religious fabric of Malay society, as well as reclaiming Malay rural dominance (Kathiramby-Wells, 247–48). To re-publish this novel as a celebration of Independence, then, is to adopt an ideological position regarding what is, or should be, organic to the nation. These types of conceptions of place can extend to geography to create what Joel S. Kahn labels a "race-space" (15), eventually mirroring the processes that resulted in such unhelpful geo-racial entities as "the Malay Archipelago" and "West Indian."

POSTCOLONIAL ECOCRITICISM

I want to suggest, finally, that looking at these regions analogously *can* be enabling. But here we must use analogous in the specific sense outlined earlier, that of taking into account its specific evolution and concrete historical context when remarking on performative similarities.

Tracing affinities between regions, while remaining aware of their emergence from very different contexts, approaches I think, a responsible methodology of comparison that considers them in an analogous fashion without reducing their unique complexities. An appropriation of Foucault works nicely, within a postcolonial, ecocritical attempt at comparison: "One finds isomorphisms . . . ," writes Foucault in *The Order of Things*, "that ignore the extreme diversity of the objects under consideration" (xi).

Although examples of isomorphisms abound, explicit comparisons of Caribbean and Southeast Asian literatures fall under the category of generally neglected transnational scholarship.[13] There are, however, a few useful models attempting similar trans-oceanic comparisons.

In *Routes and Roots*, Elizabeth DeLoughrey, provides insightful affinities, or what Édouard Glissant might term (in English translation) "succulencies of relation" between the Caribbean and the *Pacific* Islands. DeLoughrey seeks "trans-oceanic trajectories of diaspora . . . underlining the[se regions'] shared similarities in geo-pelagic relation rather than the limiting model of national frameworks" (23). To this end, she locates conceptual similarities between regions. The justification for such comparison perhaps arises from the history of U.S.-military involvement in the Caribbean and the Pacific. Indeed, in her preface she expresses an interest in "decoloniz[ing] the trajectories of U.S. militarization across the seas" (xiii).

DeLoughrey's work is thus postcolonial and in its foregrounding of geography and belonging, ecocritical. I highlight DeLoughrey's project because, interestingly, she builds much theoretical foundation for comparing/or "navigating" Caribbean and Pacific Island texts on the work of Édouard Glissant, and she does this despite Glissant opening his monumental work *Caribbean Discourse* with that declaration of Caribbean opacity or singularity: "Martinique is not a Polynesian island" (1). Though DeLoughrey never explicitly engages with this particular statement, the discourse of relational islands pervades her work.

While her study chooses to privilege the islandism of the regions over and above the continental element of the Caribbean, she demonstrates the benefits of considering the regions in their complexity, interrogating both indigeneity and diaspora through "tidalectics" (an evocative term coined by Edward Kamau Brathwaite) as an alternative epistemology to western colonialism. This is a very important thing to do, especially in the Caribbean where there exists what Maximillian Forte calls "the historical trope of anti-indigeneity" (1), where diaspora is taken as an over-arching principle despite the presence of Caribbean texts that "nativize the Caribbean landscape" (De-Loughrey 229).

DeLoughrey *does* explicitly engage with another influential metaphorical entity, however: "the repeating island," which, in Antonio Benítez-Rojo's specific usage, signifies the "meta-archipelago" that is the Caribbean. For Benítez-Rojo, the Caribbean—that marvelous place of "sociocultural fluidity," "ethnological and linguistic clamor," "historiographic turbulence," of creolization, supersyncretism, heterogeneity (3)—is *best* thought of in metaphorical terms. Benítez-Rojo goes so far as to write that "to persevere in the attempt to refer the culture of the Caribbean to geography—other than to call it a meta-archipelago—is a debilitating and scarcely productive project" (24).

There is an interesting isomorphism suggested in Benítez-Rojo's *The Repeating Island*; he declares that ". . . the Caribbean is not a common archipelago, but a meta-archipelago (an exalted quality that Hellas possessed, and the great Malay Archipelago as well), and as a meta-archipelago it has the virtue of having neither a boundary nor a center" (4). Because Benítez-Rojo's focus is strictly the Caribbean, this analogy is only a suggestion. But here we have an explicit, if brief, reference of a structural affinity between the Caribbean and the Malay, or Southeast Asian, Archipelago, as well as the Greek. How are these related? We know already how much the Grecian model informs, if circuitously, Western culture.

In Elizabeth DeLoughrey's critique of Benítez-Rojo, she examines the pernicious flip side of his concept of the repeating island exemplified by Britain (and by extension here we can add other European powers like Spain and France). DeLoughrey demonstrates how the concept of the repeating island "has ample historical precedence in British imperialism," "an older and more pernicious model of colonial island expansion" and that all it takes is a glance at the "long colonial history of mapping island spaces" to recognize a pathology of "nesomania," or "desire for islands" symptomatic of European empires (6).

In direct contrast to Benítez-Rojo's approach, which dismisses attempts to locate an original island, DeLoughrey considers, "the 'root' or originary island" of British imperialism: *England*. As DeLoughrey points out, England can only call itself an island if it suppresses Scotland and Wales. And the United Kingdom exists because of its colonial *expansion* overseas, *first*, into the territory of its immediate neighbors (e.g., Ireland) and then a rapid, fractal replication in seemingly random remote locations farther overseas such as Singapore, Jamaica, India, Guyana, Australia, Canada, the Falklands, and Belize.

In these cases, Britain is articulated as an expanding isle that is extended through the work of "transoceanic male agents of history" (DeLoughrey 26). We get, with British colonialism, the projection of imperial England's cultural topography onto other spaces in repetitive fashion. DeLoughrey argues that the colonizers repeated what they knew of Europe, so that we get so many Britains. This is also an easy point to illustrate: just think of all the New Englands there are, New York, Little Britain, Nova Scotia, New Albion, New Hebrides, New Ireland, all of the places named after Victoria. . . .

Focusing more specifically on the colonial city, Latin American theorist Ángel Rama locates a period in colonial history when Spanish conquerors "adapted themselves to a frankly rationalizing vision of an urban future, one that ordained a planned and repetitive urban landscape and also required that its inhabitants be organized to meet increasingly stringent requirements of colonization, administration, commerce, defense, and religion" (4). The cities that the colonizers left, having roots in medieval European cultures, were, in Ángel Rama's words, "organic" rather than "ordered" (3). It was only the lands of the newly conquered and colonized that would provide a "blank space" on which to construct an urban project ideally suited to the reigning social order of the day: colonial administration.

Different strategies of ordering were used for particular contexts. For example, in his contribution to a book *Theorizing the Southeast Asian City as Text*, Rajeev Patke points out that differences in policy for places as near to one another as Singapore and Malaysia had ramifications for the urban landscape that were different in each place and can be seen in the patterns of urban assimilation in each. However, in both cases, with colonization, the world is ordered according to an ideal of empire. This ordering consists of a palimpsestic new creation, just as Peter Hulme pithily states: "the gesture of discovery is also a ruse of concealment" (1).

Not all models of repetition are so epistemologically violent; Benítez-Rojo figures the Caribbean model as a non-violent, creolizing fractal expansion. While creolization is something of a clichéd term in Caribbean discourse, it is helpful to place it in juxtaposition with an isomorphism in the Malay Archipelago: nusanterism.[14]

Nusantera, the Malay word describing that East Indian archipelago, forms the root of nusanterism or, a native imaginary of the Malay Archipelago. The Malaysian poet and theorist Muhammad Haji Salleh, in his poetry and essays on literary theory, invokes this term to signify an egalitarian understanding of regional geo-history and a plea to retain a universally-applicable indigenous wisdom as resistance to the globalized consumer culture present inside the Malay Archipelago; for example, in Singapore where "the [regional] hinterland is steadily being forgotten" (Raslan 85). In its claim to an indigenous aesthetic, nusanterism might operate in a conceptually similar manner to Brathwaite's *tidalectics*, as part of a dynamic tradition. The Malay Malaysian poet, Muhammad Haji Salleh, in "Tropics," articulates a straight-forward belonging: "but the beach is the brown people's home, / their traditions / engraved by every tide" (94). When left to operate solely in its originary context, the concept risks turning from a symbolic attachment to place into a series of ethnic essentialisms. Salleh's invocation then becomes primarily an ideological battle with other indigenous values being supplanted by a generic trauma of Malay separation from place sparked by a modern cultural xenophobia. Juxtaposition against Caribbean instances of regional imaginary, however, enables a much more productive dialogue.

To interrogate nusanterism alongside Aimé Césaire's notion of *négritude* for example, or the regional imaginary expressed in his *Cahier d'un retour au pays natal*, where the persona declares, "And my original geography also: the map of the world drawn for my own use, not dyed with the arbitrary colours of men of science but with the geometry of my spilt blood, I accept . . . " (125), would be to inject a helpful and broadening discourse into interpretations of nusanterism, where one could see clearly how critiques of Césaire's négritude might apply to nusanterism, and vice-versa.

There have been recent complaints of the omission of Southeast Asia in postcolonial studies (Chua 2008). Chua Beng Huat writes that "there are substantive and conceptual issues emerging out of local Southeast Asian colonial and postcolonial experiences, which bear comparison with those of the rest of the postcolonial world" (239). Rather than belatedly insist on a Southeast Asian contribution within existing conventional frameworks, however, a responsible approach is mandated in ecocritical postcolonialism, especially the type of South-South (periphery-periphery) comparison I propose here. Removing Europe as the sole intermediary in relation avoids conventional postcolonial frameworks where both the Caribbean and Southeast Asia are cut up into discrete parts by colonial language. The Philippines is no longer linked to Cuba only through Spain; the anti-colonialism in the work of Cuba's José Martí is comparable to that of Philippine's national hero José Rizal on other grounds. Suriname and Indonesia need not be brought into comparison only because of the Dutch movement of Malay peoples, culture, and language.

The range of themes touched upon in this chapter—the Non-Aligned Movement, contemporaneous figures in anti-colonial and postcolonial struggles, nusanterism, and négritude, the pernicious effects of colonial and touristic models of the regions, the historical ironies of postcolonial conservation and environmentalism, the concepts of the meta-archipelago, tidalectics, the shared sense of architectural impermanence of these tropical regions, the role of the writer, strategies of representation of landscape, strategies of indigenization—all reveal the justification for comparison present in the enabling interrogation of analogous structures. Refusing the center-periphery dynamic, these two regions present themselves as candidates for comparison.

NOTES

1. For example, Kerstin Oloff (2009) provides a reassessment of Wilson Harris's early work from a regionalist perspective that reveals both its attractiveness for postcolonial studies and its importance as an intervention in the specific national-level discourses of its time.

2. Perhaps the most (in)famous of these attempts is represented in Jameson's "Third World Literature in the Era of Multinational Capitalism" (1986) which Aijaz Ahmad famously denounces for the presumption of attempting to articulate a cognitive theory of the third world aesthetics. See (Jameson, 1986) and (Ahmad, 1987).

3. Réda Bensmaïa's observation in *Experimental Nations* on critical responses to North African literatures is apt here: "What has long struck me was the nonchalance with which the work of these writers was analyzed. Whenever these novels were studied, they were almost invariably reduced to anthropological or cultural case studies. Their literariness was rarely taken seriously. And once they were finally integrated into the deconstructed canon of world literature, they were made to serve as tools for political or ideological agendas. This kind of reading resulted more often than not in their being reduced to mere signifiers of other signifiers, with a total disregard for what makes them literary works in and of themselves" (6).

4. Or as Neil Lazarus puts it: "'globalisation' was never the deterritorialised and geo-politically anonymous creature that neo-liberal ideology projected it as being. On the contrary, it was from the outset a political project, a consciously framed strategy designed to restructure social relations world-wide in the interests of capital" (11).

5. In one of the very few studies done that consider East and West Indies within the same intellectual frame—Murray, Boellstorff, and Robinson's "East Indies/West Indies: Comparative Archipelagoes" in *Anthropological Forum*—the related term "homologous" is invoked as the opposite descriptive nomenclature. The authors "[look] at the East Indies and West Indies, which might be seen to sit at opposite poles of anthropological inquiry, as a single unit, a unit of comparison but also one of *homology* and surprisingly parallel themes" (222, italics mine). I contest this assertion in this essay. Rather than offering remarkably parallel structures that perform differently, comparative literary studies of both regions yield, to my mind, the realization that very different cultural forces have impelled the development of works that perform similarly.

6. In its orthodox usage in the context of the biological sciences analogous structures signify appendages that perform the same function though they developed as a result of very different evolutionary pathways, for example: birds' and bats' wings.

7. See Bensmaïa's quote in n.3.

8. The word "imaginary" has resonance in other fields, but I take Martiniquan poet and theorist Édouard Glissant's definition, that is, "all the ways a culture has of perceiving and conceiving of the world" (*Poetics*, xxii).

9. Shakespeare plays on the word "cheater" which also signifies a seafaring vessel for transporting said wealth.

10. Graham Huggan complicates this position in *The Postcolonial Exotic* by arguing that successful postcolonial writers are often aware of their position within the global literary economy as "exoticised agents," and can offer a critique of their own relation to power via a "cultivated exhibitionism" (xi).

11. Being the only major novel to emerge from Belize post-independence does place certain demands on the novel as a representative work.

12. It is useful to read Ishak Haji Muhammad's *Putera Gunung Tahan* (1937) alongside Albert Memmi's *The Colonizer and the Colonized* (1957). Both identify, for example, how the profit motive of colonial empire results in mediocre officers attending positions of authority. Ishak also nicely illustrates Memmi's concept of the "Nero Complex" despite predating his theories by some twenty years.

13. Elleke Boehmer and Bart Moore-Gilbert have discussed the neglect of the transnationalist dimension within postcolonial studies. Among other issues, they implicate the tendency to focus on center–periphery (rather than periphery-periphery) dynamics.

14. Silvio Torres-Saillant argues that terms specific to Caribbean discourse is coopted by the more fashionable postcolonialism. Here, in comparing a Southeast Asian discourse with a Caribbean one, both retain their specificity in relation.

WORKS CITED

Ahmad, Aijaz. "Jameson's Rhetoric of Otherness and the 'National Allegory'." *Social Text* 17.3 (1987): 3–25. Print.

Benítez-Rojo, Antonio. *The repeating island: the Caribbean and the postmodern perspective.* Durham: Duke UP, 1997. Print.

Bensmaïa, Réda. *Experimental Nations or, the Invention of the Maghreb.* Trans. Alyson Waters. Princeton: Princeton UP, 2003. Print.

Boehmer, Elleke, and Bart Moore-Gilbert. "Introduction: Transnational Resistance in Postcolonial Studies." *Interventions: International Journal of Postcolonial Studies*, Special issue on Transnationalism 4.1 (2001): 7–21. Print.

Brathwaite, Edward Kamau. *ConVERSations with Nathaniel Mackey.* Staten Island, NY: We Press, 1999. Print.

Césaire, Aimé. *Cahier d'un retour au pays natal.* Trans. Mireille Rosello and Annie Pritchard. Newcastle upon Tyne, UK: Bloodaxe, 1995. Print.

Chakravorty, Dipesh. *Provincializing Europe: Postcolonial Thought and Historical Difference.* Princeton: Princeton UP, 2000. Print.

Chow, Rey. *Ethics after Idealism.* Bloomington: Indiana UP, 1998. Print.

———. "The Old/New Question of Comparison in Literary Studies: A Post-European Perspective," *ELH* 71 (2004): 289–311. Print.

Chua, Beng Huat. "Southeast Asia in Postcolonial Studies: an Introduction." *Postcolonial Studies* 11.3 (2008): 231–41. Print.

DeLoughrey, Elizabeth. *Routes and roots: navigating Caribbean and Pacific Island literature.* Honolulu: U of Hawai'i P, 2010. Print.

Deveson, Richard. "The Life of the Political Past." *Caribbean Women Writers.* Ed. Harold Bloom. Philadelphia: Chelsea House, 1986/1997. Print.

Donne, John. *John Donne's Poetry; Authoritative Texts, Criticism.* Ed. Arthur L. Clements. New York: Norton, 1966. Print.

Edgell, Zee. *Beka Lamb.* London: Heinemann, 1982. Print.

Forte, Maximilian C. "Extinction: The Historical Trope of Anti-Indigeneity in the Caribbean." *Issues in Caribbean Amerindian Studies* 6.4 (2004): 1–24. Print.

Foucault, Michel. *The Order of Things.* London: Tavistock, 1970. Print.

Glissant, Édouard. *Caribbean Discourse: Selected Essays.* Trans. J. Michael Dash. Charlottesville: UP of Virginia, 1999. Print.

———. *Poetics of Relation.* Trans. Betsy Wing. Ann Arbor: U of Michigan P, 1997. Print.

Grove, Richard. *Green Imperialism: Colonial Expansion, Tropical Island Edens, and the Origins of Environmentalism, 1600-1860.* Cambridge, UK: Cambridge UP, 1995. Print.

Hallward, Peter. *Absolutely Postcolonial: Writing Between the Singular and the Specific.* Manchester, UK: Manchester UP, 2001. Print.

Huggan, Graham. *The Postcolonial Exotic: Marketing the Margins.* London: Routledge, 2001. Print.

Hulme, Peter. *Colonial encounters: Europe and the native Caribbean, 1492-1797.* London: Methuen, 1986. Print.

———. "Beyond the Straits: Postcolonial Allegories of the Globe." *Postcolonial studies and beyond.* Ed. Ania Loomba, Suvir Kaul, Matti Bunzl, Antoinette Burton, and Jed Esty. Durham, NC: Duke UP, 2005. Print.

Ishak Haji Muhammad. *The Prince of Mount Tahan.* Kuala Lumpur, Malaysia: Heinemann Educational Books (Asia), 1980. Print.

Jameson, Frederic. "Third World Literature in the Era of Multinational Capitalism." *Social Text* 15.4 (1986): 65–88. Print.

Kahn, Joel. *Other Malays: Nationalism and Cosmopolitanism in the Modern Malay World.* Singapore: National U of Singapore P, 2006. Print.

Kathirithamby-Wells, J. *Nature and nation forests and development in Peninsular Malaysia.* Singapore: National U of Singapore P, 2005. Print.

Lazarus, Neil. "Postcolonial Studies after the Invasion of Iraq." *New Formations: A Journal of Culture/Theory/Politics* 59 (2006): 10–22. Print.

Lim, Shirley Geok-lin. "Postcoloniality in Southeast Asia?" *SARE: Southeast Asian Review of English* 47 (2007): 1–11. Print.

Memmi, Albert. "Mythical Portrait of the Colonized." *The Colonizer and the Colonized.* Boston: The Orion Press, 1967. Print.

Muhammad Haji Salleh. "Tropics." *The Second Tongue: an anthology of poetry from Malaysia and Singapore.* Ed. Edwin Thumboo. Singapore: Heinemann, Educational Books (Asia), 1976. Print.

Mukherjee, Upamanyu Pablo. *Postcolonial Environments: Nature Culture and the Contemporary Indian Novel in English.* London: Palgrave, 2010. Print.

Murray, David A. B., Tom Boellstorff, and Kathryn Robinson. "East Indies/West Indies: Comparative Archipelagoes." *Anthropological Forum* 16.3 (2006): 219–27. Print.

Oloff, Kerstin. "Wilson Harris and Postcolonial Theory." *Perspectives on the "Other America."* Ed. Kerstin Oloff and Michael Niblett. Amsterdam: Rodopi, 2009. Print.

Patke, Rajeev. "Benjamin in Bombay? An Asian Extrapolation." *Theorizing the Southeast Asian City As Text: Urban Landscapes, Cultural Documents and Interpretative Experiences.* Ed. Robbie B.H. Goh and Brenda S.A. Yeoh. Singapore: World Scientific, 2003. Print.

Patteson, Richard F. *Caribbean Passages: A Critical Perspective on New Fiction from the West.* London: Lynne Rienner, 1998. Print.

Raleigh, Walter. *Sir Walter Raleigh's Discoverie of Guiana.* Ed. Joyce Lorimer. Aldershot, UK: Ashgate, 2006. Print.

Rama, Ángel. "The Ordered City from *The Lettered City.*" *Foucault and Latin America : Appropriations and Deployments of Discursive Analysis.* Ed. Benigno Trigo. London: Routledge, 2002. Print.

Raslan, Karim. "The Singaporean dilemma." *Journeys Through Southeast Asia: Ceritalah 2.* Singapore: Times Books International, 2002. Print.

Rizal, José. *Noli Me Tangere.* Trans. Harold Augenbraum. New York: Penguin, 2006. Print.

Roach, Joseph. *Cities of the Dead: Circum-Atlantic Performance.* New York: Columbia UP, 1996. Print.

Robinson, Jennifer. "Comparison—Cosmopolitan or Colonial?" *Singapore Journal of Tropical Geography* 32 (2011): 125–40. Print.

Said, Edward. *Orientalism.* New York: Penguin, 2003. Print.

Shakespeare, William. "Merry Wives of Windsor." *The Works of William Shakespeare.* Ed. William George Clark, John Glover, and William Aldis Wright. Cambridge, UK: Cambridge UP, 1863-66/2009. Print.

Torres-Saillant, Silvio. *An Intellectual History of the Caribbean.* New York: Palgrave Macmillan, 2006. Print.

Chapter Five

Wai tangi, Waters of Grief, *wai ora,* Waters of Life: Rivers, Reports, and Reconciliation in Aotearoa New Zealand

Charles Dawson

Kei raro i ngā tarutaru, ko ngā tuhinga o ngā tūpuna
Beneath the herbs and plants are the writings of the ancestors
—Māori proverb (in Waitangi Tribunal, *Kō Aotearoa Tēnei* 103)

Many cultures have traditions that site water as healer, water as home. New Zealand, a skein of river-laced islands in the South Pacific, buffeted by wind, rain, and earthquake, is home to a substantial and instructive indigenous Māori tradition regarding the relationship with water. The Māori understanding of water values the dual restoration of *wai* (water) and *wairua* (the spiritual realm). The reconciliation movement between Māori and the government in New Zealand (particularly as it pertains to rivers) offers insights for ecocritics attending to the space where indigenous knowledge and negotiated policy compromises meet.

In discussing these issues it is useful to take account of the language used in the reports and subsequent government agreements within that reconciliation movement. These texts contain the explication of indigenous Māori oral traditions. Ecocritics will find the emergence of Māori concepts and metaphors as defining forces within legal settlement documents of interest. Such linguistic and legal hybridity is partly formed through presentation of evidence before the Waitangi Tribunal, a quasi-judicial body that is part of New Zealand's reconciliation process. This study focuses on aspects of the Waitangi Tribunal's functions and reports. The reports attend to the nexus of place, environment, and culture so central to ecocriticism, bridging cultures and law. Several of the Tribunal's most significant reports on specific New Zealand harbors and rivers offer insights into past and present Māori views of water. These reports have had an important impact on the process of reconciliation in a post-colonial context. In 2012, for example, one cumulative effect of one hundred and twenty-five years of Māori petitions, evidence, reports, and negotiation was a government agreement to grant the Whanganui River full legal personality, so the river can have standing. It is instructive to look at this development, for how often have ecocritics and others sought standing for the natural world in law, and how often has that been achieved?

In the pages ahead, I trace some Māori views on water, sketch out key points in New Zealand's environmental history, and then discuss the Tribunal's hearings and some aspects of its report on the Whanganui River. Tribunal hearings are sites of legal and lyrical conjunction, in which tribal evidence is presented, for example, via oratory, chant, song, woven cloaks, and the carvings in meeting houses and traditional tattoos on skin. In the hearing process, tribal tradition demonstrates enduring connections with mountains, rivers, ancestors, and allied stories. The Tribunal accords real weight to oral tradition: that evidence sits alongside legal and historical research (Phillipson). This multiplicity of voices and evidence result in a report that works with the bicultural nuances of law and lore to test government action against the principles of the nation's 1840 Treaty of Waitangi. Even though Tribunal reports have their limitations (their recommendations are generally not binding, and they are too legalistic and compromised for some), these official documents recognize two world views and try to reconcile them. That reconciliation has been central to the Tribunal's vision since its formation in 1975. These issues are live, they matter, and they have traction in political, resource-management, and legal arenas. They have also altered Pākehā (White) understandings of history, place, and story, and they use the arts and tradition to begin a process of healing (Dawson 41).

WATER RIGHTS AND RITES

Indigenous peoples' perspectives on water are of great importance: they impart understandings gleaned from centuries of habitation. These waterways are sources of survival and identity. *Wai,* or water, is a fundamental component of Māori survival on both physical and spiritual levels. As the scholar Te Ahukaramū Charles Royal observes, *Te taha wairua* can literally be translated as "the dimension of two waters," a conception that likens spirituality to water. However, it might be argued that *te taha wairua* does not mean "the spiritual plane" at all. Instead, references to *te taha wairua* might be saying that there is a fundamental dimension to all life and it takes the form of water.

Māori attend closely to the concept of *mauri*. *Mauri* is energy, health, and life essence: "the mauri is that power which permits these living things to exist within their own realm and sphere" and tracing it is one way to map river health (Barlow 83). Linda Te Aho summarizes the Māori perspective:

> Rivers are often conceptualized as living beings, ancestors with their own mauri (life force), mana (prestige and authority), and tapu (sacredness). As such they are whole and indivisible entities, not dissected into beds, banks, and water not into the tidal and non-tidal, navigable and non-navigable parts. The saying 'ko au te awa, ko te awa ko au' (I am the river and the river is me) comes from the Whanganui River, but speaks to this interconnectedness that lies at the heart of the way many Māori view the world and our waterways. ("Freshwater" 103)

On one level *wairua* translates to mean "two waters." If rivers can bind Māori and other New Zealanders, they can also divide, for that river connection has often been lived out in two different ways, and there is no certainty that water will be a unifying force in the nation's future. Colonization largely ignored the Māori view of water: Māori were often badly affected by major hydroelectricity projects on physical and spiritual levels, but they were very rarely given a say about them beforehand (Young, *Islands* 192–97). In more recent years Māori have not shirked from protest to drive the point home: for example, 1995's occupation of a lower Whanganui River site, major 2003–2004 foreshore protests, and the 2012 protest and litigation over control and ownership over freshwater foreground water rights (Keane, "Protest"; Harris; Te Awa Tupua).

In 2012, the New Zealand Māori Council claimed the government's proposed 49 percent privatization of state hydropower companies would breach Treaty rights and obligations, as the Māori right to water was not determined or protected; the Council went to the Tribunal seeking an urgent hearing. Under intense pressure from all parties, the Tribunal reported at speed: "the matters in this claim are of national importance and at the core of the Maori-Crown partnership sealed in 1840" (Waitangi Tribunal, *Wai* 2358, xviii). The Tribunal's recommendations were largely dismissed by the government. The asset sales were completed in 2014 after further testing before the Supreme Court. The Tribunal noted its numerous reports on water bodies: "found that Maori possessed their water bodies as whole and indivisible resources, in customary law and in fact. Maori did not possess only the beds of rivers or lakes; they possessed water regimes consisting of beds, banks, water, and aquatic life" (Waitangi Tribunal, *Wai*, 2358, xiii). This holistic understanding of a river as a unitary being, a *taonga*, or treasure, of the highest spiritual and physical importance, runs throughout evidence before the Tribunal: "The living relationship between the claimants and the resources is an integral part of its status as a taonga" (Wheen and Ruru, "Environmental" 101). Thus it is that the nation's longest river, the Waikato, is bound to the name of the Waikato-Tainui tribe along its banks, and the Māori name for the South Island is *Te Wai Pounamu*, the jade waters, while the Māori name for the country, *Aotearoa*, land of the long white cloud, indicates the land is blessed with rain.

In Aotearoa/New Zealand, the river being of great mana or prestige for Māori is the longfin eel. Like the salmon for indigenous peoples of the north Pacific Rim, the eel sustains Māori river communities, visitors, and story. The longfin eel, found only in New Zealand, is the world's largest eel; its ecological status is fragile in this age of hydropower and pollution. One recent report noted it was on "a slow path to extinction" (McDowall 142–236; Young, Woven 178–90; Parliamentary Commissioner 6; Keane). Displacement of species begets displacement of knowledge. Since the 1880s, Māori eel weirs have been destroyed for navigation: the traditions of trap, weir, and eel-pot construction also declined. The large weirs of the Whanganui River, for example, were substantial structures: in 1880 there were some 350; within thirty years most had gone (Waitangi Tribunal, Whanganui 70; Keane, "Eels"). R. M. McDowall observes there were many instances of "irrational paranoia" where preservation of eel populations was given no priority, in comparison with endless official promotion of waters for (introduced) trout fishing (156). In the twentieth century, massive and often innovative hydropower dams and diversions harnessed the energy of major rivers crucial for Māori, even as they bequeathed the nation relatively clean energy generation.

Some Māori traditions regard eels as the offspring of *Te Ihorangi*, "a personified form of rain" forced to leave their drought-afflicted celestial home to find new sites on earth (McDowall 158). For Māori, the eel becomes the basis for a "multidimensional continuum" of tradition, ritual, and myth, accounts that catalyze remembered (or revived) customary practices. This continuum of story, history, and tradition has been represented in much New Zealand literature, film, and theater in recent decades, and it is central evidence before the Waitangi Tribunal: "What is clear is that there is among Māori a distinct desire to perpetuate their traditional freshwater fisheries, in much the same way as they seek to perpetuate their language" (McDowall 156, 789).

FROM BIRD LAND TO AOTEAROA TO NEW ZEALAND

New Zealand is the planet's last major temperate landmass to be settled by humans. Isolation has marked both its ecology and its resident humans' zest for adaptability and endurance. Until Polynesian settlers arrived about 1250, "these were, above all, islands of birds" (Young, *Islands* 18; Irwin). Human footfall "initiated the great assault upon the New Zealand fauna." Major burnings of rainforest for plantations, and the introduction of rats, dogs, and hunting, rendered nearly forty bird species extinct (Anderson 28–29). But through readjustment for sheer survival, the Polynesian voyagers named the plants, birds, and sites of the new islands. In their adaptations to a colder climate, and the refinement of *whakapapa* (a genealogy that encompasses plants, animals, deities, and humans), they shaped a form of sustainability for survival and became Māori, "normal," developing a wealth of chants and oral traditions sourced from and in the *whenua*, a word that means both land and placenta (Park "Whenua" 206; Waitangi Tribunal, *Kō Aotearoa* 104; Mead 269; Young, *Islands* 38–56). European settlement terraformed the land: "By 1844 the colonial surveyor's grid was sweeping over the coastal plains like a mechanical tide" as other imported survival modes were laid out and land acquisition accelerated (Park, Uruora 40, Boast). In the late 1860s, as the New Zealand Wars peaked, "Maori felt they had lost control of their own land" (Ward 148). From the 1870s more bush was burnt, European populations grew, and 85 percent of wetlands were drained (Park, "Swamps" 151). Pastoral commerce was largely removed from Māori control, while cities began to spread. The remnants of rainforest that survived often did so because they were acquired from Māori for "scenic" purposes (Young, *Islands* 124).

THE TREATY OF WAITANGI, THE WAITANGI TRIBUNAL, AND A REVEALED HISTORY

In 1975, New Zealand novelist Witi Ihimaera observed:

> There are two cultural maps of my country, the Maori and the Pakeha. The Pakeha map is dominant, its contours so firmly established that all New Zealanders, including Maori, are shaped by it. The Maori map has eroded and, although its emotional landscape is still to all intents and purposes intact, it has been unable to shape all New Zealanders, including Pakeha. (215)

Scott Slovic urges the ecocritic to make "tales of experience and hope . . . audible in the halls of power" (142). With 15 percent of New Zealand's 4.5 million people self-identifying as Māori, and with a noticeable Māori presence in our mixed-member proportional representation system of parliament, those voices are often heard. Tribunal reports are worthy of attention from ecocritics who are alive to the force of story and history in the face of a rapidly changing world. Some of those changes, including a wider *Pākehā* awareness of Māori perspectives, have been significant. Increased public tension over race relations, and an invigorated confrontation with the past recalibrated notions of belonging through the 1970s and 1980s: "Māori artists, writers, and musicians released the spirit and drive of the enduring debate into their works" (Harris 91). Patrick Evans describes the impact of that change:

> The significance of these times for the dominant literary and intellectual culture, the sense that we were being turned through 180 degrees and forced to address our past, to gather and weigh and assess it before filing it away, is most fully revealed in the increasing manufacture of its nonfiction. (26)

In this period Māori were moving to a more public stage, while the literary form was "another adaptation of European tools to a long struggle against cultural disappearance conducted through written appeal, symbolic action, and increasingly in the 1970s, direct protest" (Williams 209, 210). Meanwhile opposition was growing in New Zealand, as environmental groups began to push for a say in, for example, massive water infrastructure and hydropower developments: "Undoubtedly, the Tribunal's focus on issues of environmental degradation struck a chord with large sections of the Pakeh community and helped solidify public acceptance of the Tribunal's role" (Hamer 5). The Treaty of Waitangi 1840 was central to the reports and protests around land loss.

> Historian Alan Ward describes some key elements of the Treaty of Waitangi:
> The Crown officials concluded a treaty which, in return for Maori recognition of the Queen's sovereignty (kawanatanga), confirmed and guaranteed to them the possession (rangatiratanga) of their lands, forests, and fisheries (their taonga). But British settlement was taken by the Crown as a given, and Maori were expected to make way for it. If they did not, the treaty guarantees were watered down or set aside. (167)

Both established principles of colonial law and the Treaty provided "the pre-existing rights of Māori would be respected" but in fact Treaty breaches were common (Waitangi Tribunal, *Whanganui* 339). By 1975, when Māori urbanization compounded a growing distress and anger, it was imperative that government act. One response was the creation of a quasi-judicial body—the Waitangi Tribunal—that would inquire into claims from any Māori regarding Crown actions or omissions claimed to breach the principles of the Treaty of Waitangi.

TRIBUNAL ROLE

By 1985, "the Maori demand for justice was overwhelming, and Parliament could not turn its back" (Ward 30). In what has been called "an act of enormous courage and insight . . . without parallel in the world," the Tribunal was now given the power to inquire into claims of Treaty breach since 1840 (Snedden 183, 184). The way was opened for the Tribunal to consider the entire colonial period, land confiscations, and Native Land Court determinations. Since 1985, the Waitangi Tribunal has contributed to "a subtle revolution" in attitude at the middle ground of compromise, accommodation, and reconciliation (Hamer 5; Sharp 142). Nonfiction on Māori issues grew as the history of colonial "legalised rapacity" was laid bare (Ward 167; Gibbons 86-94). As Paul Hamer puts it, "the Tribunal was at the forefront of a nation coming painfully to terms with its past for the first time. The process was always going to be fraught with difficulties" (6). Over the next generation, millions of pages of Māori and Crown records and panoply of oral and written Māori tradition underwent a new form of scrutiny: "The new history was no less than a history of injustice; and it was largely the republication of an older Māori history" (Sharp 4). There are up to twenty Tribunal members. Members have expertise in areas such as Māori culture and tradition, law, history, business, resource-management planning, and governance. In an inquiry, panels of Māori and non-Māori experts hear evidence and produce a report. These reports attend to the Māori and English language versions of the Treaty of Waitangi and Treaty principles, within a statutory context, to share something of the Māori evidence presented to the panel.

TRIBUNAL HEARINGS

Over a series of hearings (often lasting weeks), claimant witnesses and their families, the Tribunal panel and staff, lawyers, media, and the wider community listen as oral traditions and written evidence are presented. Tribunal hearings are intense events. Grievances held across generations are aired in on the traditional territories of the claimants. Ancient and new songs and many memories are relayed inside the carved meeting houses, those gathering spaces that are seen as a living embodiment of ancestral presence. Claimants may appreciate the chance to tell their own stories to their community. But the preparation and hearings are grueling and place great demands on all involved. Parties face the reality of compromise within a constrained Treaty settlement process that, as both government and Māori acknowledge, can never fully compensate for what has been lost. There remains some strong *Pākehā* opposition to Treaty settlements or affirmative action aimed at addressing inequities of the past. Yet the Tribunal has said, "It is neither a privilege nor racist that a people should be able to retain what they have possessed" (Waitangi Tribunal, *Whanganui* 339). And the Māori voice is diverse: many Māori argue they had no real say in designing the Treaty settlement process; for some the process damages Māori identity, Treaty educators have argued "there are significant weaknesses in the tribunal process" (Consedine 115). The Crown-Māori power imbalance is still marked, and the constitutional change many Māori seek is not on offer in Treaty settlements; moreover, despite millions of words and many encouraging speeches, racism is still an issue and Māori still figure disproportionately in most negative socio-economic indicators (Bargh; Sykes; Poata Smith). Governments, meanwhile, do not always welcome the Tribunal's reports: "Its future is constantly under review in the politics of the nation" and can promote reform that dilutes environmental protections (Ward, "Foreword," Palmer).

TRIBUNAL REPORTS

Tribunal reports attend to the interaction between cultures, languages, and the disciplines of law and history. The reports are partial records of the evidence placed by witnesses before the Tribunal. They are always partial: "Such poor justice is done to the power of the spoken word by writing about it" (Brody). The Tribunal can work with Māori songs, orations, physical site visits, and eyewitness accounts of historical events: "These different forms of evidence contribute to a complex, layered interpretation of historical claims" that goes beyond text alone (Phillipson 41). The reports have been regarded as "an invaluable resource on traditional beliefs and practices, and on Crown policies and practices and the law since 1840, and their effects on Māori people" (Wheen and Ruru 98–99). The Māori understanding of environment and place is central: here *whakapapa*, the genealogical web of connection between humans and the natural world, is realized and given meaning. It is this intersection that will often appeal to the ecocritic as the use of Māori language and ideas from the evidence of elders and experts brings an indigenous worldview into the legal arena. Assessing their early impact, Andrew Sharp suggested "one of the most notable features of the Tribunal's work was the art and sophistication with which it listened to and relayed a Māori version of history to a wider audience" (Sharp, *Justice* 4).

When the Tribunal finally issues its reports, after poring over many thousands of pages and weeks of evidence, those reports are recommendatory only. Crown-Māori negotiations (which do not require a tribunal report) result in Treaty settlements that involve financial and cultural redress and important Crown apologies. Settlements have become increasingly innovative in response to changing social attitudes and developments in co-governance design (an example is the July 2014 settlement passed by every political party in Parliament regarding Urewera National Park and its Tūhoe guardians). The next decades will see Māori taking more of a role in the economic and conservation systems that have historically excluded them (Waitangi

Tribunal, *Kō Aotearoa*). The Waitangi Tribunal has revealed that history to the system itself: "a culture's most precious places are not necessarily visible to the eye" (Park, *Uruora* 20).

WATER, IDENTITY, AND THE WHANGANUI RIVER

"Water is at the heart of the earliest Waitangi Tribunal reports" (Te Aho 2012, 102). From the 1983 report on the Motunui coast which faced massive pollution due to a synthetic fuels plant, the 1984 report on the Kaituna River that prevented sewage outfall to the river, the 1985 report regarding a steel mill planned for Manukau harbour, or the major later reports on many North Island rivers, claimants have been clear that the health of a waterbody links with the health of the people; the currents are *taonga*, treasures held in trust of inestimable value, for those Māori who maintain connections.

Tribunal reports quote those who appear before it to try to bring the (hi)stories of origins, connection, and continuities forth. In the Te Ika Whenua Rivers Report, for example, claimant witness Wiremu McAuley speaks of the special prayers recited before fishing or eeling, and of the healing properties of the water, which was used for rituals by whanau (families) far from home:

> The water from the puna wai [water of the spring] of a whanau [family] is considered a taonga to that whanau as it carries the Mauri [life force] of that particular whanau. Of course all the waters of the puna wai find their way into the river and thereby join with the Mauri of the river. In essence then the very spiritual being of every whanau is part of the river . . . . In this sense the river is more than a taonga[;] it is the people themselves. (Waitangi Tribunal, *Ika* 13)

This inextricable link between *hapū* (sub-tribe, clan), *iwi* (tribe), and water occurs again and again in evidence before the Tribunal and is testament to those who have borne claims through generations and managed to keep their traditions alive, something that is harder and harder to do in the cities (Mataamua and Temara).

> Reading the landscape is like collage, interweaving the patterns of ecology and the fragments of history with footprints of the personal journey. The journey, in time as well as in space, plays no small part.
> —Geoff Park, *Ngā Uruora* (16)

The Tribunal has observed that each river case must be looked at according to its circumstances (Waitangi Tribunal, *Whanganui* 338). But it has consistently found that Māori did not consent to the removal of rights to water. Those rights were at the fore in the country's highest court, when state hydropower assets sales were further challenged by Māori (Supreme Court, 2013). Māori rights continue to offer up real challenges to a business-as-usual or majority-rules approach. For New Zealand, hydropower asset sales are set to continue, but the Supreme Court has required the process to adhere to Treaty principles, which include partnership, equity, and good faith.

If the reports of the 1980s and early 1990s on rivers from the Tribunal were instructive, significant, and often too challenging for full political assent, the largest of the Tribunal river reports—that concerning the country's longest navigable river, the Whanganui—would nevertheless ring in some of the strongest calls for recognition of Māori water rights and rites at the fundamental level of national identity: "by coming to understand the real import of this claim, the Tribunal report suggests, New Zealand society will be re-founded in terms of indigeneity" (Johnson 102). The report would also track one of the country's oldest litigation streams, for Māori of the Awa, of the River, have sought justice for over 125 years. The tribal struggle for recognition of both their rights and attendant obligations reiterates the river as a political and cultural site. This is how the Waitangi Tribunal's interim report on the river claim opens:

> Rarely has a Māori river claim been so persistently maintained as that of the Whanganui people. Uniquely in the annals of Maori settlement, the country's longest navigable river is home to just one iwi, the Athaunui-a-Paparangi. It has been described as the aortic artery, the central bloodline of that one heart. (Waitangi Tribunal, *Whanganui*, Appx. 4 379)

From *Te Ātihaunui-a-Pāpārangi*, the Māori tribal network woven along the length of the Whanganui River, there comes a saying:

> E rere kau mai te Awanui
> Mai i te Kahui Maunga ki Tangaroa
> Ko au te Awa, ko te Awa ko au.
> The Great River flows
> From the Mountains to the Sea
> I am the River, and the River is me. (Tūtohu 3)

River elder Matiu Mareikura told the Waitangi Tribunal:

> [The river] . . . is our life cord, not just because it's water—but because it's sacred water to us. Our people go to the river to cleanse themselves, they go to the river to pray, and they go to the river to wash. They go to the river for everything leads to back to the river. And the river in return suffices all our needs. Without the river we really would be nothing. (Waitangi Tribunal, *Whanganui* 57)

On the surface the Whanganui seems a "fallen" river: its headwaters are diverted for hydro-electricity generation and for decades the river had been the repository for the raw sewage from the larger river towns, a desecration that violates Māori practice in several ways. The Whanganui is not a "virtual river," according to Richard White's definition of the Columbia River in the United States; but his assertion—"[w]hat has failed is our relationship with the river"—rings true on the Whanganui (White 62, 106). There may be a chance to heal that relationship. This is why learning from the Māori is so vital. Most importantly, the diversion has not entirely silenced the tribal stories and songs of the Awa, nor the capacity for intercultural dialogue.

An annual journey for Whanganui river Māori, the *Tira Hoe Waka*, or journey by canoe, becomes a means of reconnection with the stories and songs (ancient and modern) of the Awa, the river. The journey honors the past and present and reminds one there are several ways of knowing water, of hearing as "the River speaks" (National Film Unit). Each annual journey is another thread in a weave of stories (Waitangi Tribunal, *Te Kāhui Maunga* 1367; *Te Awa Tupua*). Tariana Turia, who is of the Whanganui River and has held various Ministerial roles in two governments, has spoken of the *Tira Hoe Waka* as "a vehicle for our survival." It offers strength to contend with modern life and the lingering legacies of colonization. So for many Māori of the river, its waters retain some capacity for healing.

The Tribunal's Whanganui River Report conveys an enormous amount of emotion, fact, imagery, and intensity. Its power resides in its capacity to research and reveal the river's hidden stories, to bring the facts of the past to light and assess them with care, and to reassure the wider public that stronger Māori engagement and control over resources will be of benefit to the River and the community. Yet like many Tribunal reports, its key findings were initially rejected by the government of the day in 1999. But in 2012, when a significant new Treaty relationship framework was proposed for the Whanganui River, things had moved on, (hi)stories had percolated, and people recognized that the time for the river had come, that it needed healing, and that relationships needed healing as well.

A RIVER GIVEN STANDING

In August 2014, Whanganui River tribes and the New Zealand government signed a deed for settling the tribe's Treaty of Waitangi river claim. The Deed and its 2012 framework was developed around two key principles centered on *Te Awa Tupua*, the river of prestige:

> Te Awa Tupua mai i te Kahui Maunga ki Tangaroa—an integrated, indivisible view of Te Awa Tupua in both biophysical and metaphysical terms from the mountains to the sea; and

Ko au te awa, ko te awa ko au—the health and wellbeing of the Whanganui
River is intrinsically interconnected with the health and wellbeing of the people. (Tūtohu 5)

The river will be recognized as a complete entity "incorporating its tributar-
ies and all its physical and metaphysical elements," rather than by imposing
the English common law tradition of division of the river into water, bank,
and adjoining land (Tūtohu 5). For the first time in Western legal history a
river attains legal personality:

> The creation of a legal personality for the River is intended to:
> reflect the Whanganui Iwi view that the River is a living entity in its own right
> and is incapable of being 'owned' in an absolute sense; and
> enable the River to have legal standing in its own right. (Tūtohu 10)

The proposal is to appoint two River Guardians (one appointed by
government, another by the River *iwi*) as the human face of the river, and as
key leaders in river management. The goal is to support and improve the
health of the River, the *mana* of the River itself, from which flows the *mana*
of the *iwi*. Significantly, the guardians are not accountable to their appoint-
ers, but to the river itself. The river is leader, guide, and teacher. Ecocritics
may appreciate the Deed's use of Māori knowledge and concepts from river
lore as organizing metaphors for river management. Here Māori knowledge
is an operational and regulatory force in law and in legal language. The
image of the supplejack vine, used to make eel traps, is also the binding and
supporting concept for those people who support the river guardians, while
the description of the framework for guiding the operation of the new agreement draws again on the metaphor and practice of eels as sustaining forms of
life force and mana:

> Te Pā Auroa—The Broad Eel Weir
> He pā kaha kua hangaia kia toitū ahakoa ngā waipuke o te ngahuru, o te
> makariri me te kōanga
> The broad eel weir built to withstand the autumn, winter and spring floods
> The name of the Te Awa Tupua framework symbolises an extensive, well-
> constructed framework for Te Awa Tupua that is fit for purpose, enduring and
> the responsibility of all. (Ruruku 4)

Indigenous language has found its way into New Zealand resource management laws since the late 1980s, but is only just beginning to have a deeper
impact. In 2014, then, the legal language is beginning to become bicultural at
a functional level; these instruments have a cultural nuance that offers challenge and opportunity, in part because it offers restitution at the level of
regulatory expression and implementation.

This Whanganui deal was hammered out while other Māori tribes pushed for ownership of the waters of other rivers may have provided extra momentum for the parties. Laura Hardcastle suggests the deal was "a compromise to prevent iwi from gaining ownership" and that environmental safeguards are not as strong as they might be. But as David Young points out, "'The Whanganui is about the stuff that opens doors,' [Maori] say. The doors are still opening" (Young, *Woven* 177). The river has functioned as a crossroads "of cultures and discourses" for Māori and for Pākehā (Newton 173). Like most healing journeys that path is uncertain, but it can be a healing journey:

> As ever, the rivers of the earth have time on their side; it is we who are the losers. Given time, nature will heal itself, with or without human beings. Our healing lies in recognising that the Whanganui's current problems are really about this country's race relations. While never forgetting that the often-unintended impositions of colonialism will not clear in the first spring freshes, there is also the opportunity to learn again from both nature and from one another. Much of this process will be through the telling of story. (Young, Woven 257–58)

Gaston Bachelard suggests "water invites us on an imaginary journey" (183). Māori oral tradition is full of tales of transformation, from the trickster Māui hauling up a giant fish that became the North Island to the eels making their passage from the stars to the rivers on earth (Royal; McDowall 158). Despite change and loss, there are continuities. The remarkable opportunity here in New Zealand is that Māori have patiently offered other New Zealanders clues to "think like a river," for the benefit of all (Worster 331). In 2012, a wide-ranging collective of water users suggested we have a once in a generation chance to give effect to a new consensus on managing our freshwater (Land and Water Forum). Subsequent discord over river claims puts this opportunity at risk (McCulloch). In part any fragile consensus will turn on the capacity to continue to nourish a shared vision of the restoration and reconciliation of river and story.

ACKNOWLEDGMENTS

I would like to thank David Young and Paul Hamer for their comments on an earlier draft of this chapter and feedback from the editors and W. H. New, as well as support from Gerrard Albert and Che Wilson. Any errors or omissions remain my own. I would like to extend my thanks to the Atihaunui-ā-Paparangi tribe of the Whanganui River for sharing their lessons in river wisdom, compassion, determination, and patience with my family for decades. My father Martin worked with them on the river claim and the people introduced me to the depths of story and current they tend. This chapter does not interrogate or explicate all historical or legal issues; nor does it canvas a wide range of literary works from an ecocritical perspective (that is done very well by Alex Calder in his 2011 book *The Settler's Plot: How Stories Take Place in New Zealand*). But it is worth noting Stephen Turner's assertion that "[t]he New Zealand literary tradition . . . has been primarily concerned with place" (27). Attention to the connection with nature or place is evident in much New Zealand poetry, for example (Brown 10). New Zealand literature today eschews easy labels, remaining alive to "contestation and differentiation" (Sturm xviii). The Tribunal's reports are available online and provide a relatively accessible source of material to readers who wish to follow up the very brief summaries here.

WORKS CITED

Aho, Linda. "Ngā Whakataunga Waimāori. Freshwater Settlements." *Treaty of Waitangi Settlements*. Ed. Nicola Wheen and Janine Hayward. Wellington, New Zealand: Bridget Williams, 2012. 102–13. Print.

Bachelard, Gaston. *Water and Dreams: An Essay on the Imagination of Matter*. Trans. E.R. Farrell. Dallas, TX: The Dallas Institute for Humanities and Culture, 1983. Print.

Bargh, Maria. "The Post-settlement World (So Far): Impacts for Māori." *Treaty of Waitangi Settlements*. Ed. Nicola Wheen and Janine Hayward. Wellington, New Zealand: Bridget Williams, 2012.166–81. Print.

Barlow, Cleve. "Tikanga Whakairo." *Key Concepts in Maori Culture*. Auckland: Oxford UP, 1991. Repr. 1994. Print.

Boast, Richard. "Te tango whenua—Māori land alienation—Land tenure and alienation." *Te Ara: the Encyclopedia of New Zealand*. Web. 22 September 2014.

Brody, Hugh. *Maps and Dreams*. Vancouver, Canada: Douglas & McIntyre, 1981. Print.

Brown, James, ed. *The Nature of Things. Poems from the New Zealand Landscape*. Nelson, New Zealand: Craig Potton, 2005. Print.

Calder, Alex. *The Settler's Plot: How Stories Take Place in New Zealand*. Auckland, New Zealand: Auckland UP, 2011. Print.

Consedine,, Robert and Joanna. *Healing our History. The Challenge of the Treaty of Waitangi*. Auckland: Penguin, 2005. Print.

Dawson, Charles. "Enduring Rivers of Light: waters of memory, Aotearoa & Aniwaniwa." *AJE: Australasian Journal of Ecocriticism and Cultural Ecology* 3 (2013–2014): 31–44. Print.

Evans, Patrick. *The Long Forgetting. Post-Colonial Literatures in New Zealand*. Christchurch, New Zealand: Canterbury UP, 2007. Print.

Gibbons, Peter. "Non-fiction." *The Oxford History of New Zealand Literature in English*. 2nd ed. Auckland, New Zealand: Oxford UP, 1998. 31–118. Print.

Hamer, Paul. "A Quarter-century of the Waitangi Tribunal. Responding to the Challenge." *The Waitangi Tribunal/Te Roopu Whakamana i te Tiriti o Waitangi*. Ed. Janine Hayward and Nicola R. Wheen. Wellington, New Zealand: Bridget Williams, 2004. 3–14. Print.

Harris, Aroha. *Hīkoi: Forty Years of Māori Protest*. Wellington, New Zealand: Huia, 2004. Print.

Hardcastle, Laura. "Turbulent times: speculations about the Whanganui River's position as a legal entity." *Māori Law Review* Feb 2014. Web. 5 August 2014.

Ihimaera, Witi. "Why I Write." *Writers on Writing: An Anthology*. Ed. Robert Neale. Auckland, New Zealand: Oxford UP, 1992. 215–16. Repr. from WLWE 14.1 (1975). Print.

Irwin, Geoff, and Carl Walrond. "When was New Zealand first settled?—Genealogical dating." *Te Ara—The Encyclopedia of New Zealand*. 22 September 2012. Web. 12 March 2012.

Johnson, Miranda. "Burdens of Belonging. Indigeneity and the Re-Founding of Aotearoa New Zealand." *New Zealand Journal of History* 45.1 (2011): 102–12. Print.

Keane, Basil. "Te hopu tuna—eeling - Pā tuna—eel weirs." *Te Ara—The Encyclopedia of New Zealand.* Web. 22 September 2012.

Keane, Basil. 'Ngā rōpū tautohetohe—Māori protest movements—Historic Māori protest." *Te Ara—The Encyclopedia of New Zealand.* Web. 13 July 2012.

Land and Water Forum. *Land and Water Forum Publishes Third and Final Report.* Press release. 15 November 2012. Wellington, New Zealand: Land and Water Forum Trust, 2012. Print.

Mataamua, Rangi, and Pou Te Rangiua Temara. "Tūhoe and the environment. The impact of the Tūhoe diaspora on the Tūhoe environment." *Māori and the Environment: Kaitiaki.* Ed. R. Selby, P. Moore and M. Mulholland. Wellington, New Zealand: Huia, 2010. Print.

McCulloch, Alison. "Muddied Waters." *Werewolf.* Web. 6 September 2012.

McDowall, R. M. Ikawai. *Freshwater Fishes in Māori Culture and Economy.* Christchurch, New Zealand: Canterbury UP, 2012. Print.

Mead, Hirini Moko. *Tīkanga Māori. Living by Māori Values.* Wellington, New Zealand: Huia, 2003. Print.

National Film Unit. *The Legend of the Whanganui River.* Wellington: NFU, 1953. Film.

Newton, John. *The Double Rainbow.* Wellington, New Zealand: Victoria UP, 2011. Print.

Palmer, Rt. Hon. Sir Geoffrey. "The Resource Management Act: How we got it and what changes are being made to it." Paper delivered to the Resource Management Law Association Conference, New Plymouth, New Zealand. Web. 27 Sept 2013.

Park, Geoff. "'Swamps which might doubtless easily be drained.' Swamp drainage and its impact on the indigenous." *Environmental Histories of New Zealand.* Ed. Eric Pawson and Tom Brooking. Melbourne, Australia: Oxford UP, 2002. 151–65. Print.

———. *Ngā Uruora: The Groves of Life. History and Ecology in a New Zealand Landscape.* Wellington, New Zealand: Victoria UP, 1995. Print.

Parliamentary Commissioner for the Environment. *On a Pathway to Extinction? An investigation into the status and management of the longfin eel.* Web. 13 September 2013.

Phillipson, Grant. "Talking and Writing History. Evidence Before the Waitangi Tribunal." *The Waitangi Tribunal/Te Roopu Whakamana i te Tiriti o Waitangi.* Ed. Janine Hayward and Nicola R. Wheen. Wellington, New Zealand: Bridget Williams, 2004. 41–52. Print.

Poata Smith, E.S. Te Ahu. "The Changing Contours of Maori Identity and the Treaty Settlement Process." *The Waitangi Tribunal/Te Roopu Whakamana i te Tiriti o Waitangi.* Ed. Janine Hayward and Nicola R. Wheen. Wellington, New Zealand: Bridget Williams, 2004. 168–83. Print.

Pocock, J.G.A. "Tangata Whenua and Enlightenment Anthropology." *NZJH* 26. 1 (1992): 28–53. Reprinted in *The Shaping of History.* Ed. Judith Binney. Wellington, New Zealand: Bridget Williams, 2001. 38–61. Print

Royal, Te Ahukaramū Charles. "Tangaroa: the sea. Water as the source of life." *Te Ara—The Encyclopedia of New Zealand.* Web. 22 September 2012.

Ruruku Whakatupua Te Mana o te Awa Tupua. *Agreement between Whanganui River Iwi and the Crown.* Web. 5 August 2014.

Sharp, Andrew. *Justice and the Māori. The Philosophy and Practice of Māori Claims in New Zealand since the 1970s.* 2d ed. Auckland, New Zealand: Oxford UP, 1997. Print.

———. "The Trajectory of the Waitangi Tribunal." *The Waitangi Tribunal/Te Roopu Whakamana i te Tiriti o Waitangi.* Ed. Janine Hayward and Nicola R. Wheen. Wellington, New Zealand: Bridget Williams, 2004. 195–206. Print

Slovic, Scott. "'There's Something About Your Voice I Cannot Hear': Environmental Literature, Public Policy, and Ecocriticism." *Going Away to Think. Engagement, Retreat, and Ecocritical Responsibility.* Reno: U Nevada P, 2008. 134–42. Print.

Sturm, Terry. "Introduction." The Oxford History of New Zealand Literature in English. 2nd ed. Auckland, New Zealand: Oxford UP, 1998. ix-xix. Print.

Supreme Court of New Zealand. New Zealand Maori Council (and ors) v The Attorney-General (and ors). Supreme Court of New Zealand. SC 98/2012 [2013] NZSC 6. Web. 27 February 2012.

Sykes, Annette. "Bruce Jesson Memorial Lecture 2010." 13 July 2011. Web. 22 September 2012.

Te Awa Tupua: Voices From the River. Dir. Paora Joseph. Rongomai Productions, 2014. Film.

Tūtohu Whakatupua. *Agreement between the Crown and Whanganui River Iwi.* Wellington: *Office of Treaty Settlements.* Web. 30 August 2012.

Tipa, Gail, and Laurel Tierney, et al. *A Cultural Health Index for Streams and Waterways: Indicators for recognising and expressing Māori values. A Toolkit for Nationwide Use.* Wellington, New Zealand: Ministry for the Environment, 2006. Web. 22 September 2012.

Turia, Hon. Tariana. "Speech to Healing our Spirit: Worldwide Conference." Sixth gathering. Honolulu, Hawai'i. 9 September 2010. Conference.

Turner, Stephen. "Settlement as forgetting." *Quicksands: Foundational Histories in Australia and Aotearoa New Zealand.* Ed. Klaus Neumann, Nicholas Thomas, and Hilary Ericksen. Sydney, Australia: U of New South Wales P, 1999. Print.

Waitangi Tribunal. *Te Ika Whenua Rivers Report.* Wellington, New Zealand: Waitangi Tribunal, 1998. Print.

———. *The Whanganui River Report.* Wellington, New Zealand: Waitangi Tribunal, 1999. Print.

———. *Kō Aotearoa Tēnei: A Report into Claims Concerning New Zealand Law and Policy Affecting Māori Culture and Identity.* Wellington, New Zealand: Waitangi Tribunal, 2011. Print.

———. *Te Kāhui Maunga: The National Park District Inquiry Report: Pre-publication.* Vol. 3. Wellington, New Zealand: Waitangi Tribunal, 2012. Print.

———. *Wai 2358. The Stage 1 Report on the National Freshwater and Geothermal Resources Claim.* Wellington, New Zealand: Waitangi Tribunal, 2012. Print.

Ward, Alan. *An Unsettled History: Treaty Claims in New Zealand Today.* Wellington, New Zealand: Bridget Williams, 1999. Print.

———. Foreword. *The Waitangi Tribunal/Te Roopu Whakamana i te Tiriti o Waitangi.* Ed. Janine Hayward and Nicola R. Wheen. Wellington, New Zealand: Bridget Williams Books, 2004. xi. Print.

Wheen, Nicola and Jacinta Ruru. "The Environmental Reports." *The Waitangi Tribunal/Te Roopu Whakamana i te Tiriti o Waitangi.* Ed. Janine Hayward and Nicola R. Wheen. Wellington, New Zealand: Bridget Williams, 2004. 97–112. Print.

White, Richard. *The Organic Machine: The Remaking of the Columbia River.* Toronto, Canada: HarperCollins, 1995. Print.

Williams, Mark. "The Long Maori Renaissance." *Other Renaissances.* Ed. Zhou Gang, Sander Gilman, and Brenda Deen Schildgen. London, UK: Palgrave, 2006. 207–26. Print.

Worster, Donald. "Thinking Like A River." *The Wealth of Nature: Environmental History and the Ecological Imagination.* New York: Oxford UP, 1993. 123–34. Print.

Young, David. *Woven by Water: Histories of the Whanganui River.* Wellington, New Zealand: Huia, 1998. Print.

———. *Our Islands Our Selves: A History of Conservation in New Zealand.* Dunedin, New Zealand: Otago UP, 2004. Print.

———. "Whanganui River Tribes." *Te Ara—The Encyclopedia of New Zealand.* Web. 4 December 2012.

Chapter Six

Fish, Coconuts, and Ocean People: Nuclear Violations of Oceania's "Earthly Design"

Dina El Dessouky

Conventional discourses of economic development situate most Pacific Island nations within the Global South. Nuclear industrialization—an economic model which now includes nuclear energy development but which has its roots in the environmentally destructive colonial project of nuclear weapons testing—is one among several prominent paths towards capitalist "development" which Euro-American financial discourses have historically proposed for the Pacific region.

As Arif Dirlik points out, the 1980 and 1983 Brandt Commission reports are largely responsible for the popularization of the term "Global South"; Dirlik notes that these reports "advocated large infusions of capital from the North to the South to enable their modernization" (13), which suggests that the term emerged from within a neocolonial, capitalist, and North-dominant perspective. The perspective informing the initial creation of the Global South concept places the "sponsor" countries of the Global North (i.e., the United States, Britain, France, and Japan) in the coveted position on the top rung of the international modernization hierarchy by virtue of their capital advantages. In doing so, it asserts neocolonial tendencies through its suggestion that those of the South should aspire to be more like their Northern counterparts by adopting Northern economic models.

While the onset of nuclear weapons testing campaigns in the Pacific predates the coinage and rise of the term "Global South" itself, Euro-American proposals for nuclear testing in Pacific colonies and territories stem from logic similar to the Brandt Commission report, in that both encouraged the South to follow the model of the North in order to attain both modernization and financial progress. For example, a 1990 interview conducted with Ma'ohi (Indigenous French Polynesian) poet Henri Hiro by fellow Ma'ohi writer Rai Chaze suggests that France linked nuclear testing with economic opportunity and modernization in an effort to persuade the Ma'ohi people to accept its presence in French Polynesia. Hiro notes:

> De Gaulle arrived in 1960 and made a speech announcing to us his "personal determination" to open Polynesia to "civilization." He said that this opening couldn't be accomplished without building an airport. I don't think that he mentioned, however, that after the airport would come the CEP; later we learned that Fangataufa and Moruroa had been handed over to France for nuclear-testing installations. After that came the military, more easy money, massive immigration, and so forth. Nuclear testing contaminated everything. (qtd. in Chaze 79)

While this excerpt from the interview presents economic and infrastructural changes to French Polynesia that global capitalist economics would deem beneficial—namely the construction of both an international airport and the Centre d'Expérimentation du Pacifique (CEP), both major new employers within French Polynesia during France's 1966–1996 nuclear testing campaign—the last sentence reflects these changes not as positive influences, but, ultimately, as part of the larger destructive process of nuclearization. This is partly because Hiro speaks from an Indigenous Ma'ohi point of view, one which treats cultural and environmental health as greater priorities than capitalist economic progress. When Hiro notes that the nuclear campaign "contaminated everything," he is describing the chain reaction of socio-economic shifts affiliated with the CEP, an institution some locals have given the moniker of the "social bomb" (Greenpeace International 5). He evaluates the airport and the CEP not simply as advancements for French Polynesia, but more in terms of their responsibility for the toxic health and environmental effects and social, cultural, and economic fallout that accompanied thirty years of nuclear tests. This fallout included an influx of French "métropolitain" workers and military personnel threatening to solidify French language and cultural hierarchies, an increased reliance on foreign goods paid by heavy import taxes, and a mass immigration of Ma'ohi who would leave their families, ancestral villages, and islands for what they saw as the lucrative opportunities of French Polynesia's capital city, Papeete.

Hiro and the other Pacific Island voices in this chapter document and contest nuclear testing through their activism and literature. The focus of these voices is primarily on how nuclear testing negatively impacts both Oceanic place and Oceanic being rather than on lauding any of its potential economic or infrastructural benefits. Such a focus also helps us rethink the appropriateness of the Global South framework for Oceania. In proposing the insufficiency of the Global South as a framework for Oceania, it is important to first acknowledge the potential promise of applying the concept to Oceania. Many Pacific Island communities share histories or ongoing contexts of Euro-American (neo)colonialism with other constituencies of the Global South, and thus might find potential in the Global South as a common network of identification through which Oceanic peoples and communities living in other forefronts of the nuclear industry might organize transnational coalitions or regional antinuclear solidarity movements. Therefore, while the Global South is no longer simply an economic model and has taken on "changing usages" each implying their own unique agendas (Dirlik 13), the origins of the term in the U.S. and Western European institutions, epistemologies, and governmental systems may disqualify it as the most empowering choice for Oceania's primarily Indigenous populations. The Northern-, capitalist-, democracy-centric connotations of the term imply that it is limited in its capacity to aid homegrown regional decolonization efforts.

I agree with Arif Dirlik's statement that "Alliances that cut across developmental maps may be an urgent requirement of the times" (12), but instead propose the concept of Oceania—rather than the Global South—as a better fitting "alternative global alliance" (12) in which communities in Oceania may participate. Oceania's prime feature—the boundless ocean—offers not only a more appropriate regional framework, but also the potential for relational and planetary coalitions.

In order to demonstrate that the economics-based Global South framework is limiting to and thus not ideal for Oceania, the remainder of this essay argues that Indigenous literature and oral testimony from Oceania posit nuclear weapons testing as a simultaneously local, regional, and planetary concern. Each text articulates notions of what Kanaka Maoli/Native Hawaiian writer Haunani-Kay Trask's 1985 poem "People of the Earth" calls "earthly design." This concept posits culture and place as co-constitutive of Indigenous identity, and suggests that the environmental integrity of a specific place is foundational to the cultural identity and survival of the Indigenous peoples of that place. Together with Trask, the oral testimonies of Maʻohi (Indigenous French Polynesian) nuclear test site workers and the poetic commentary of Kathy Jetnil-Kijiner (Marshallese), Bobby Holcomb (Kanaka Maoli), and Teresia Teaiwa (I-Kiribati/Banaban) collectively reveal that the vast ocean itself provides the basis for the place-based identities of many of Oceania's Indigenous communities. Trask's poem refers to those basing their identity in Oceania's expansive depths as "ocean people," and suggests that the devastating environmental effects of twentieth century nuclear testing by global superpowers such as the United States and France in the Pacific on its islands and waters had the tandem effect of destabilizing the cultural identities of Indigenous Pacific Islanders. To further illustrate the anchoring of Oceanic place-based identities in both the land and sea, I examine the frequent references that Oceanic texts addressing nuclear testing make to fish and coconuts—the two common and beloved Indigenous food sources in Oceania. I read Oceanic references to the contamination of fish and coconuts as acknowledgments of nuclear testing's broad, layered reach—most potently at the local and regional level, but ultimately, throughout the planet. While textual accounts of the irradiation of Oceania's vast swaths of air, water, and life are demoralizing and can be read as apocalyptic, their attention to the connective, symbiotic, borderless, and boundless qualities of the Pacific Ocean underscore Oceania as a relational place which necessitates reciprocity and cooperation. I close the essay with a proposal to read Oceanic texts and the histories they record cyclically. To read cyclically is not to limit Oceanic writings and histories to linear progressions, and, furthermore, to recognize that as forms of Indigenous knowledge production, they refuse to give up on the cyclical and regenerative capacity of life processes.

"OCEAN PEOPLE"

Here is the opening of Haunani-Kay Trask's 1985 poem, "People of the Earth": "culture and place / together / made of us / what we are . . . //" (140). In the poem, the Kanaka Maoli/Native Hawaiian poet, activist, and scholar explores an experience common to many Indigenous communities worldwide—the compromise of specific, place-based Indigenous identities and cultures at the hands of the global military-industrial complex. With her line, "ocean people grow ghostlike in the wake of nuclear seas" (140), Trask's poem points to the impact and legacy of nuclear weapons testing, which the United States spearheaded in 1946 in the Marshall Islands. The United States went on to test in the Marshall Islands through 1958, and was followed by other countries such as Britain and France. France's nuclear testing campaign in the Tuamotu Atolls lasted from 1966 to 1996.

The U.S. and French governments suggested to their constituents back home that atolls such as Bikini, Rongelap, Enewetak, Kwajalein, Moruroa, and Fangataufa were so remote or so sparsely inhabited that they were ideal sites for nuclear testing. Marshallese poet Kathy Jetnil-Kijiner exposes the dehumanizing implications of this nuclear colonialist rationale in her poem "History Project," which discusses the discoveries and reflections her poem's narrator made as a teenager while researching a history project on U.S. nuclear testing in her Native islands. She writes: "I sift through political jargon / tables of nuclear weapons / with names like Operation Bravo / Crossroads / and Ivy / quotes from generals like / 9,000 people are out there. Who cares?" (Jetnil-Kijiner "History Project"). In addition to indicating the United States' lack of care for the health and livelihood of at least 9,000 people—and countless human generations to come—comments such as these by U.S. military generals reveal that the U.S. government also saw Native environments themselves as—to borrow Val Plumwood's critical vocabulary—"unused or empty areas of rational deficit" (53).

Trask's use of the term "ocean people" to refer to the Indigenous communities most affected by such nuclear testing contests this degrading Western colonial rationale. The specific "earthly design" of Indigenous Oceanic people is one which situates identity in both the local land mass of each specific people, in addition to being something that stretches beyond the limits of each island to incorporate the entire, borderless ocean realm as a basis for that identity. The term highlights time-honored Oceanic perspectives of the diverse human and non-human life-forms and elements of Oceania as interdependent and significant. The term "ocean people" echoes Tongan scholar Epeli Hauʻofa's statement that historically, for the peoples of Oceania:

> Their universe comprised not only land surfaces, but the surrounding ocean as far as they could traverse and exploit it, the underworld with its fire-controlling and earth-shaking denizens, and the heavens above with their hierarchies of powerful gods and named stars and constellations that people could count on to guide their ways across the seas. Their world was anything but tiny. ("Our Sea" 90)

Instead of stating directly to her reader that she is referring to Marshallese or Maʻohi ("French" Polynesian) peoples, Trask foregrounds the common, place-based bonds that connect Indigenous Pacific Islanders with their ancestral lands, the living salt water expanse surrounding their islands, and the spirit worlds lying within and beyond the ocean world. Thus, "ocean people" is also a strategic, relational term that encourages a re-examination of the links between Indigenous communities throughout Oceania. Similar to Hauʻofa's use of the term, "kakai mei tahi" ("people from the sea") ("Our Sea" 91), "ocean people" is an identifier which counteracts colonial projections of these groups as separate and isolated, opting instead to emphasize their far-reaching and fluid connections.

While the idea of a person's body being of "Earthly Design" likely meant little to the U.S. military generals whom Jetnil-Kijiner's "History Project" critiques, it is foundational to the identities of the "ocean people" referenced in Trask's, "People of the Earth." Read within the poem's larger context, the first line of Trask's poem—"culture and place / together / made of us / what we are //" ("People" 140)—suggests that "Earthly Design" is the common Indigenous understanding that a person's specific culture and place are co-constitutive of her Indigenous identity. However, Trask's poem reminds us that the "Earthly Design" of all people and life is vulnerable to the devastating impact of nuclear explosions. Her poem notes that the radioactive contamination wrought by nuclear testing interrupts Indigenous Oceanic lifeways, resulting in the dispossession of ocean people from their ancestral ocean places. The fallout from nuclear testing compromises the ocean person's freedom to "fish / and chant / long voyages" (140) within his ocean. The choice of global superpowers to experiment with their radioactive arsenals leaves ocean people to "grow ghostlike / in the wake / of nuclear seas" (140-41).

FISH

The effects of nuclear testing proved most devastating closest to the detonation sites in the Marshall Islands and French Polynesia. Test site workers and residents living within close range of nuclear test sites received the worst and most direct exposure to radioactive fallout. While the French government delineated certain "zones" on the Moruroa and Fangutaufa atolls as off-limits for fishing or bathing to each site's many Ma'ohi and French site workers, the government's designation of a small zone on each atoll in which nuclear fallout would remain contained was grossly inaccurate. As Ruta, a Ma'ohi worker at Moruroa interviewed by Greenpeace noted, "the wind and fish didn't respect which zones were forbidden and contaminated" (13). Ruta's comment situates the circulatory environment of Moruroa atoll as part of Oceania's larger system of fluid, amorphous, multi-layered, and largely unknown proportions and interactions, one in which rigid nation-state legal and zoning practices are inappropriate and futile. Ruta's point was scientifically proven during an August 1968 study at the University of Baja California, where scientists found high levels of radioactivity in several fish populations along Mexico's Baja Peninsula, which they traced to French testing (Greenpeace International 18). In addition to irradiated fish turning up thousands of miles away, Tupou, a Ma'ohi worker on Fangataufa, noted in his Greenpeace interview that "after each underground explosion there was a sort of tidal wave that washed over Fangataufa and a few days later thousands of stinking dead fish washed up on the shore" (39). Since mountains of dead fish frequently covering beaches is not something that goes unnoticed, especially by people whose lives revolve around the sea, the carnage of sea life surfacing onto beaches became well known throughout French Polynesia, prompting the Kanaka Maoli artist, Bobby Holcomb, who lived on the French Polynesian island of Huahine, to compose the song, "Bikini People." The song's chorus reminds French Polynesians of all backgrounds, most of whom live in Tahiti that "Tahiti e, a'e / Moruroa's not so far away (auwē!) / and fish are dying around us everyday" (Holcomb). Holcomb's "Bikini People" song affirms his Oceanic awareness that the islands of French Polynesia/Te Ao Ma'ohi are not so distant and isolated from one another after all, but rather, are linked through the ocean elements of winds and currents, the latter of which transport contaminated fish directly from nuclear test sites.

Two series of interviews and studies conducted with French Polynesia's test site workers—one published by Greenpeace in 1990 and the other published by Dutch researchers Pieter de Vries and Han Seur in 1997—demonstrate countless devastating cases of ciguatera poisoning and cancer in workers who consumed contaminated fish. In addition, interviews with the Ma'ohi workers reveal the dilemma they faced when warned not to eat fish from the waters of Moruroa and Fangataufa. Oscar Temaru, a former worker at the French nuclear headquarters in French Polynesia and current territorial president of French Polynesia, explains this dilemma: "It's very difficult for Tahitians to stop eating fish when it has been our daily diet for centuries. Fish has always been good food—why should it suddenly be bad?" (qtd. in Greenpeace International 28). Other workers often made statements like "I ate fish in Moruroa because there we had Western food every day and after some time my body would demand fish to eat" and "Tahitians eat fish because your body craves it" (de Vries and Seur 52). Onno, who began working on Moruroa at the age of thirteen and was interviewed by de Vries and Seur, sums up the relationship between Ma'ohi and fish when he states, "Fish is our life, our body, you cannot withhold that from us" (53). Irradiated fish, then, poses a threat not only to Ma'ohi physical health, but also to the existence of Ma'ohi cultural identity.

COCONUTS

Like fish, coconuts have also been a core staple of Indigenous Oceanic well-being and sustenance for centuries, making them central components of Indigenous Oceanic identity. In his 1988 essay, Ma'ohi poet Turo Raapoto notes that the cultural identifier of *Ma'ohi*, has at its etymological root the word *'ohi*, which we can interpret as a sprout which has secured "autonomy of life . . . all the while being linked to the mother stem" (5). The reo maohi language's links between the Ma'ohi person and the 'ohi plant metaphor suggests that when indigenous plants such as coconut trees are unable in their earliest growth stages to secure this autonomy of life due to nuclear fallout, it is not only a reflection of testing's threats to plant populations, but also a warning sign that the survival of Ma'ohi people is also at stake. As wild, fruit bearing plants growing in atoll and barrier reef islands, coconut trees absorb and concentrate radiation at extremely high rates because, as Greenpeace notes, "where soils are lacking in minerals (for example in the sandy non-absorbent soils of coral atolls), caesium remains far more available," and results in higher contamination of plants (16). After finding lethal levels of strontium-90 in well water, Lawrence Livermore Laboratory scientists told Bikini Islanders in 1977 to limit their coconut intake to only one a day, which resulted in Bikini Islanders' forced reliance on food and water from the U.S.

sources (Firth 31). Stuart Firth's reference to this study on coconuts in his 1987 book *Nuclear Playground* (31) inspired I-Kiribati poet/scholar Teresia Teaiwa's haunting spoken-word poem "Bad Coconuts," which warns: "an apple a day, keeps the doctor away / but a coconut a day will kill you / if you live on Moruroa / if you visit Fangataufa / return to Enewetak / resettle Bikini / a coconut a day/ will kill you" (Figiel and Teaiwa). On each of these islands, a coconut or breadfruit a day will kill you because, as Firth states, "cesium-137, chemically similar to potassium, enters the food chain by traveling from the soil into fruits such as breadfruit and pandanus and then into the muscles of the body where it lodges, producing a risk of cancer; strontium-90 is also found in breadfruit and pandanus, and can cause leukemia" (36).

In addition to causing cancer among those who received more direct exposure to nuclear testing, eating contaminated fish or coconuts threatened "increased incidence of sterility . . . miscarriage and still births and genetic defects in children" (Greenpeace International 19). Jetnil-Kijiner's poem "History Project" asks its audience to connect "our sleepy coconut trees / our sagging breadfruit trees" of the Marshall Islands to local "first-hand accounts of what we call / jelly babies / tiny beings with no bones / skin, red tomatoes / the miscarriages gone unspoken." Jetnil-Kijiner's work here links multi-generational birth defects and traumas to the stunted growth of coconut and breadfruit trees, trees which in a non-nuclear environment, might have formed fecund, nourishing groves rather than refusing to bear fruit altogether. The poem likens the failure of the fruit trees to ripen to the failure of Marshallese babies to develop in a healthy manner, let alone thrive. Both Teaiwa and Jetnil-Kijiner demonstrate how the nuclear debilitation of local plant life and foods such as coconuts also destabilizes Indigenous livelihoods and reproductive cycles. In that sense, they reveal that nuclear testing qualifies not only as an environmental threat, but also as a potentially genocidal science experiment that the United States and France conducted against the Indigenous peoples of Oceania.

CYCLICAL READING AND OCEANIC TEXTS

The enormous reach and endurance of nuclear fallout is devastating, which makes it challenging to read optimism into Oceanic engagements with nuclear testing. Poems such as "History Project" make no attempt to sugar-coat nuclear testing's destructiveness. When the teenage narrator of Jetnil-Kijiner's poem researches U.S. nuclear testing in her homeland (the Marshall Islands), she discovers that testing has come at an immense human and environmental health cost for her islands. The poem concludes with the following lines:

so I finish my project / graph my people's death by cancer and canned food diabetes / on flow charts / in 3D / gluestick my ancestors' voice onto a poster-board I bought from office max . . . and at the top I spraypainted in bold stenciled yellow / FOR THE GOOD OF MANKIND / and entered it in the school district wide competition called / history day / my parents were quietly proud/ and so was my teacher / and when the three balding white judges finally came around to my project / one of them looked at it and said yea / but it wasn't really / for the good of mankind, though/ was it? / and I lost. (Jetnil-Kijiner "History Project")

When read linearly and only once, texts such as "History Project" can leave the reader with the heavy, final sense that nuclear testing has doomed the Pacific to an eternity of mutations and cancers. That the narrator states that she "lost" the "history day" competition because the judges fail to understand her historical critique adds to the sense that in addition to a history of U.S. nuclear colonialism, the discipline of history itself has marginalized the Pacific.

However, through a cyclical reading of "History Project"—one which returns to its opening lines and privileges cyclical processes over linear progressions—the poem suggests that Oceania continues to push onward in a state of rising consciousness instead of succumbing to the potentially paralyzing effects of the nuclear era. The opening lines of the poem, which read, "at fifteen I decide / to do my history project on nuclear testing in the Marshall Islands / time to learn my own history, I decide" (Jetnil-Kijiner "History Project") remind us that the narrator represents a generation born of survivors—not simply victims—as well as a new generation of Indigenous Pacific youth who are not afraid to confront European and American legacies of nuclear colonialism in the Pacific. While her history project's title, "FOR THE GOOD OF MANKIND" functions in part as a cynical echoing of the narrator's earlier discovery of a U.S. army general's own words to her ancestors (Jetnil-Kijiner "History Project"), it does not indicate her surrender to the nuclear legacy or to historical dogma. Rather, the narrator's investigative history project is a refusal of futility, one which pays homage to the experiences of the ancestral voices and place of the Marshall Islands, and through confrontation and storytelling, begins the process of healing place and people. The project not only sustains Marshallese culture in spite of the forces of historical amnesia and cultural erasure which the poem suggests were tools of U.S. nuclear colonialism, but it also can be read as a gift made on behalf of humanity. The narrator's history project has the courage to face the devastation wrought by nuclear testing—which has deeply wounded her islands, people, and ocean—and thus makes a convincing case for this historical, cultural, and ecological mistake never to be repeated.

Another principle of cyclical reading—the assessment of an author's work within the larger body of her writing—also confirms Jetnil-Kijiner's efforts to fortify the bonds between Marshallese culture and place which may have loosened as a result of nuclear colonialism. Jetnil-Kijiner's "Tell Them," serves as a message from the Marshallese to the continental U.S. populace that, "we don't want to leave / we've never wanted to leave / and that we / are nothing without our islands." While this poem's main goal is arguably to raise awareness about the threat of displacement that global warming and rising sea levels pose to people of low-lying atolls such as the Marshall Islands, the poem's line "we've never wanted to leave" (Jetnil-Kijiner "Tell Them") recalls the first wave of Marshallese forced migration from atolls such as Rongelap after the United States tested its nuclear weapons there, rendering the land and lagoon too toxic to sustain life. The United States deemed Rongelap safe for human habitation in 1957 and gave displaced Rongelapese clearance to resettle their atoll despite radiation levels remaining dangerously high there (Firth 40). Thus, this poem attests to the ability—and will—of Indigenous Oceanic peoples to return to their ancestral islands at the same time that it expresses the determination not to repeat this history of forced migration in light of new colonial forces. In this sense, Jetnil-Kijiner's work suggests that despite colonial practices which destabilize Oceanic culture by first targeting place, Marshallese people can persist as ocean people because they proclaim "we / are nothing without our islands" ("Tell Them"). This proclamation is indicative of what Trask would call "Earthly Design," a concept which might seem tenuous in an era of mounting Indigenous displacement and environmental destruction, but which remains stable because of its grounding in life's cyclical processes and the planet's regenerative capacities.

Cyclicality, then, is a critical component of the Indigenous Oceanic place-based knowledge that informs each of the literary texts here. To illustrate this concept, I turn to Winona LaDuke's explanation of how in Anishinaabeg and Cree traditions, a respect for the cyclicality of life processes works in tandem with a commitment to reciprocity in order to create "minobimaatisiiwin" (the "good life," or, "continuous rebirth") (79). She states:

> Two tenets are essential to this paradigm: cyclical thinking and reciprocal relations and responsibilities to the Earth and creation. Cyclical thinking, common to most indigenous or land-based cultures and values systems is an understanding . . . that the world flows in cycles. Within this understanding is a clear sense of birth and rebirth and a knowledge that what one does today will affect one in the future, on the return. A second concept, reciprocal relations, defines responsibilities and ways of relating between humans and the ecosystem. Simply stated, the resources of the economic system, are recognized as animate and, as such, gifts from the Creator. Within that context, one could not take life without a reciprocal offering. (80)

While LaDuke presents a nuanced cultural example, she applies it to her broader argument on important components of Indigenous approaches to society and environment, in general. Her comment that so-called "resources" are understood as animate forces resonates with Trask's statement that people must cease seeing place and the nonhuman entities that inhabit it as "resources" but instead recall that non/humans are equal relations existing alongside one another within a circle of life (Eros and Power 182–83). Guided by the understanding that rebirth and return is an eventual outcome within the recurring flows of life processes—one whose success hinges upon the integral actions of each being during her life—cyclicality brings an enormous amount of responsibility into all thought and action; the challenges implicit in this work, however, anticipate the privileges of a continued existence. The Oceanic literatures here—as engaged Indigenous literatures struggling against their communities' displacements, and reckoning their own roles within such displacements—remain conscious of and invested in this steadfastness to place, and thus, attempt to craft the relational discourses surrounding their community futures with the utmost care for place, for its wellness, and for their own healing.

CONCLUSION: OCEANIA'S POTENTIAL

Collectively, the fish commentary of Bobby Holcomb, Oscar Temaru, and various Ma'ohi test site workers, along with the coconut poetics of Teresia Teiawa and Kathy Jetnil-Kijiner, recall Haunani-Kay Trask's poetic concept of "earthly design." They suggest that fish, coconuts, and the cultural processes of cultivating, blessing, gathering, and preparing these place-based food sources have enabled thousands of years of existence of Oceania's Indigenous communities, and have made peoples such as the Ma'ohi and the Marshallese the distinct "ocean people" that they are. From the perspective of an ocean person, access to healthy fish and coconuts from their ancestral island places affirms their peoples' "earthly design." Thus, for Ma'ohi or Marshallese people to be suddenly unable to safely sustain themselves with ancestral foods and instead, be forced to rely on imported food of foreign occupiers that is preserved in metal tins compromises the reciprocal relationship of "culture and place together" that makes Ma'ohi and Marshallese people "ocean people."

However, Oceania's poets also offer strategies for healing and maintaining the bonds between Oceanic cultures and places—and thus restoring earthly design—despite the continuation of nuclear and other colonialisms. For Indigenous Oceanic peoples, the reciprocal offerings that are part of healthy, balanced cycles can come in the form of narrative, story, poem, or voice. Within the contexts of forced removal from place—such as nuclear colonialism—literature informed by the teachings of place becomes a critical medium through which committed connections with ancestral place can be maintained, and the duties of reciprocity to that place that nourishes one's identity and existence can also be fulfilled. In this sense, narrative—in oral or written form—proves to be a critical resource for survival that centers the Indigenous person in her origins, even in dire situations of displacement. Chicana writer, Cherríe Moraga, proposes writing as a healing practice that can help Indigenous peoples struggling with displacements recover, mend, remember and 're-member' severed physical relationships with ancestral place. She affirms: "I write to remember. / To make rite (ceremony) to remember. / It is my right to remember" (Moraga 378). The literatures that I analyze in this study proclaim their continuity with place; they actively hold space in anticipation of the eventual return of their people to their ancestral homelands and seas. They offer to island place—and those of island place, living, dead, or unborn—in words, story, emotion, and consciousness what they may at other times and contexts offer in a tangible form or action.

The fluid and connective nature of oceans illustrates that place should not be solely an "Indigenous" concern or its stewardship a specifically "Indigenous" responsibility, even if that has traditionally been the case. Similarly, significant threats to place, like nuclear industrialization (nuclear testing and nuclear energy infrastructure), are more than just an Indigenous concern, even if Oceania, its land/sea places, and its Indigenous peoples have historically been in the front lines of nuclear testing. Nuclear industrialization is a grave planetary matter, as the focus of Indigenous Oceanic voices in this essay on fish, coconuts, and the layered, circulatory, and borderless environment of the ocean asserts. The continuing spillage of radioactive material from the Fukushima Dai-ichi nuclear plant into the Pacific Ocean since Japan's March 2011 earthquake and tsunami threatens life and cultures throughout the planet, and calls into question the value of economic development models—such as nuclear energy—linked to traditional Global South rhetoric. Despite the unifying potentials of a transnational term like Global South, its economic and institutional connotations overlook the nonhuman elements and lifeforms that are so much a part of each place in which Indigenous identities base themselves, and which ultimately, sustain non-Indigenous lives as well.

Instead, "Oceania" holds more promise both as a regional identifier for those living in and around the ocean more commonly known as the Pacific, and as a basis for problem-solving and solidarity within and beyond the region. To illustrate this, I close with the scholarship of Tongan writer Epeli Hau'ofa, which suggests that humans living in Oceania and beyond—native and non-native people included—must acknowledge their mutual origins in and responsibilities towards the world's oceans. Hau'ofa notes that, "All of us in Oceania today, whether indigenous or otherwise, can truly assert that the sea is our single common heritage" ("The Ocean" 54). Hauʻofa directs his comments first towards Indigenous inhabitants of Oceania, and then towards non-Indigenous members of Oceania; he ties both groups to the sea that they share, even if the extent and depth of their cultural relationships to that sea have historically differed. He continues to posit the ocean as a coalition space when he states:

> In a metaphorical sense the ocean that has been our waterway to each other should also be our route to the rest of the world. Our most important role should be that of custodians of the ocean; as such we must reach out to similar people elsewhere in the common task of protecting the seas for the general welfare of all living things. ("The Ocean" 55)

Hauʻofa's work further implies that through the ocean, humans might establish alliances cutting across not only ethnic, national, or regional boundaries, but also possibly, across species boundaries. The suggestion that the ocean offers great coalitional potential is not an invitation to appropriate or homogenize Oceania and its specific places and cultures, or to deny that Oceania has disproportionately borne the brunt of the nuclear era's transgressions. Instead, Hauʻofa offers a model of Oceania as a headquarters for collaborative stewardship, one which emerges from within Native/Oceanic contexts and worldviews of cyclicality and reciprocal relations, but does not bar non-Native or non-Oceanic peoples from partaking in this commitment in a spirit of common destiny.

WORKS CITED

Chaze, Rai. "The Source: An Interview with Henri Hiro." *Vārua Tupu: New Writing from French Polynesia*. Ed. Alexander Dale Mawyer, Kareva Mateata-Allain, and Frank Stewart. Honolulu: U of Hawaiʻi P, 2006. 71–81. Print.

de Vries, Pieter and Han Seur. *Moruroa and Us: Polynesians' Experiences during Thirty Years of Nuclear Testing in the French Pacific*. Lyon, France: Centre de Documentation et de Recherche sur la Paix et les Conflits, 1997. Print.

Dirlik, Arif. "Global South: Predicament and Promise." *The Global South* 1.1 (2007): 12–23. Web. 15 March 2014.

Bagchi, Alaknanda. "Conflicting Nationalisms: The Voice of the Subaltern in Mahasweta Devi's Bashai Tudu." *Tulsa Studies in Women's Literature* 15.1 (1996): 41–50. Print.

Figiel, Sia, and Teresia Teaiwa. "Bad Coconuts." *Terenesia*. Hawaiʻi Dub Music and ʻElepaio Press, 2000. CD.

Firth, Stewart. *Nuclear Playground*. Honolulu: U of Hawaiʻi Press, 1987. Print.

Greenpeace International. *Moruroa—Place of the Great Secret—Testimonies:Witnesses of French Nuclear Testing in the South Pacific*. Auckland, New Zealand: Paradigm, 1990. Print.

Hauʻofa, Epeli. "Our Sea of Islands." *Asia/Pacific as Space of Cultural Production*. Ed. Arif Dirlik and Rob Wilson. Durham and London: Duke UP, 1995. 86–101. Print.

———. "The Ocean in Us." *We Are the Ocean: Selected Works*. Honolulu: U of Hawaiʻi P, 2008. 41–59. Print.

122   *Dina El Dessouky*

Holcomb, Bobby. "Bikini People." *Ohipa.* Recording Details Unavailable, 1983. CD. February 2013.
Jetnil-Kijiner, Kathy. "History Project." YouTube. 19 April 2012. Web.
———. "Tell Them." *Iep Jeltok: a basket of poetry and writing from Kathy Jetnil-Kijiner.* WordPress.com 13 April 2011. Web. 12 February 2013.
LaDuke, Winona. *The Winona LaDuke Reader: A Collection of Essential Writings.* Stillwater, MN: Voyageur, 2002. Print.
Moraga, Cherrie. "The (W)rite to Remember: Indigena as Scribe 2004–2005 (an excerpt)." *A Companion to Latina/o Studies.* Ed. Juan Flores and Renato Rosaldo. Malden, MA: Blackwell, 2007. 376–89. Print.
Plumwood, Val. *Feminism and the Mastery of Nature.* London and New York: Routledge, 1993. Print.
Raapoto, Turo A. "Maohi: on Being Tahitian." *French Polynesia: A Book of Selected Readings.* Ed. Ron Crocombe and Nancy J. Pollock. Suva: Institute of Pacific Studies of University of the South Pacific, 1988. 3–7. Print.
Trask, Haunani-Kay. "People of the Earth." *Mālama: Hawaiian Land and Water.* Ed. Dana Naone Hall. Honolulu: Bamboo Ridge, 1985. 140–41. Print.
———. *Eros and Power: the Promise of Feminist Theory.* Philadelphia: U of Pennsylvania P, 1986. Print.

Chapter Seven

Intimate Kinships: Who Speaks for Nature and Who Listens When Nature Speaks for Herself?

Benay Blend

"I never felt alone or afraid up there in the hills," writes Leslie Marmon Silko (Laguna Pueblo) in her memoir *The Turquoise Ledge* (2011). "The 'humma-hah' stories describe the conversations as coyotes, crows, and buzzards used to have with human beings" (45). In this passage, Silko notes what she has in common with the languages of other living things, specifically the land or biotic community of her immediate place. Throughout her work, she conveys in detail the notion of "what can be known without words," demonstrating here a growing sense that "the clouds and river also have their ways of communication" (45). Weaving nature and culture, event and place together into a whole, she notes that the Pueblo people have long attached certain stories to specific location: "the Pueblo people and the land and the stories are inseparable" ("Introduction," *Yellow Woman* 14). As she immerses herself in other living things and the land, Silko illustrates the "persistence of place and place-attachments" (Heise 41) that Ursula Heise claims has formed the basis of ecocritical research since it emerged as a prominent field of environmental thinking in the mid-1990s. Most importantly, Silko's words convey the following question posed by ecocritics: Who gives voice to nature and who listens when nature speaks for itself?

Speaking also with the natural world, such writers as Linda Hogan (Chickasaw), Joy Harjo (Muskogee), and Louise Erdrich (Ojibwe), among others, explore creative alternatives to alienation from nature. All turn to what Heise describes as a kind of "situated knowledge" (30), an intimate relationship with local nature that includes the material landscape and all its creatures. Shortly after moving to an old ranch house in the Tucson Mountains, Silko writes, she used physical immersion in the local landscape to integrate herself into her new surroundings. "Before long," she says, "the desert terrain and all its wonderful beings and even the weather won my heart" (*Turquoise Ledge* 81). For Native people, the land and the stories are inseparable. Because the job of storytellers, Linda Hogan explains, is to "witness" ("Two Lives" 241), her words suggest a place of "crossed beginnings" where the natural world finds its voice (*Book of Medicine* 28). Echoing indigenous people's deeply rooted and close relationship to their surroundings, ecocritical scholarship has traditionally emphasized local places as sources of identity. This "situated knowledge" (30), as Heise puts it, arises from a people's sense of place in which they live and serves as a possible site of resistance to globalization, although Heise's own 2008 book has suggested that there is a need to combine local and global sensibilities. In the environmental literature of the Global South, the privileging of the local persists, perhaps because globalizing processes are typically the processes of colonial exploitation. Native American literature, which continues to celebrate intimate relationships between human beings and the local land, even in the twenty-first century, is an important contribution to the anti-colonial, anti-globalist trends in Global South literature more generally.

In the end, writes Linda Hogan, all discussion is a dialogue with nature (*Sweet Breathing* xiv). According to Joni Adamson, however, most ecocritics have focused on writers such as Henry David Thoreau and Edward Abbey, who recognize the holistic aspect of all things, but sometimes ignore that aspect of nature that includes contested land where humans have manipulated the environment for centuries. By contrast, Navajo poet Luci Tapahonso understands that separation of self and culture from nature has caused serious environmental crises. Attuned to an alternative wisdom, she understands that spiritual regeneration is only possible as part of a community that includes the material landscape. "In the Americas," she asserts, "the sacred surrounds us, no matter how damaged or changed a place may appear to be" (*A Radiant Curve* 145). Having seen her people marginalized and impoverished due to degradation of their environment, she chooses to focus on that very landscape still replete with meaning for Navajo people.

In addition to cultural survival, Linda Hogan attributes her personal sustenance to a relationship with the land, a connection that transcends the mind/body and human/nature divisions inherent in Western thinking (*Woman* 58). Leslie Silko, too, understands the alienation and separation in inherent the Euro-American view that assumes nature and non-humans are the Other, ripe for conquer and control. Rather than taking for granted their right to consume and destroy, Silko writes, tribal people have long noticed with dismay the "devastating impact human activity can have on plants and animals." Offering instead a model for communities in harmony with nature, Silko describes a culture in which words and stories illustrate the reality of the peoples' sense of the place in which they live. "Without the plants, the insects, and the animals," she says, "human beings living . . . [in the desert] cannot survive." Accordingly, she attributes "stories about humans and animals intermarrying, and clans that bind humans to animals and plants through a whole complex of duties" to personal survival as well as that of all species (Silko, "Yellow Woman" 69). In her novels, memoir, and non-fiction essays, Silko draws on this invisible layer of meaning that covers the physical landscape and binds Laguna history and culture to the arid land.

In contrast, according to Vandana Shiva, many feminists embrace dualistic logic as a means to gain parity with men (Shiva and Mies, "Introduction" 5). As victims of this paradigm, Native Americans have long understood that capitalism exploits not only the land but also those associated with it—women and others, in particular. Paula Gunn Allen agrees: "Believing that our mother, the beloved earth," constitutes only "inert matter is destructive to yourself" ("The Woman I Love is a Planet" 80). Given the prominence of female imagery in Native American society—Changing Woman, for example, who lives at the center of the earth and "reminds us," says Tapahonso, that "she is our mother and that all comes to live as she breaths" (*A Radiant Curve* 145)—it follows that gender is allied not only with the natural world but also with oral tradition and a sense of place.

According to Hogan, women are the natural leaders in the environmental struggle. Because she feels that women have an "ancient bond" (*Sweet Breathing* xiv) with nature, both are aligned as primary players in her texts. "The green world has long been the province of women," Hogan claims, not because women are more biologically able to understand it, but because it is "synonymous with feminine wisdom" (*Sweet Breathing* xi). As part of a continuous cycle, Hogan looks back to her powerful female, hunter/gatherer ancestors whose knowledge of plants held "healing powers, sacred purposes . . . green intelligence, . . . and power" (*Sweet Breathing* xi).

As offspring of Mother Earth, Native women writers also emphasize the importance of place as a way of resisting the most pernicious forms of globalization. For example, Leslie Silko's work describes an alternative type of community reliant on the historical and personal knowing of people who have lived for long periods of time in a specific place. Labeling the land with the words and stories of her people, Silko's writing infuses every feature of contemporary landscape with the Pueblo people's sense of place in which they live. Despite five centuries of oppression and dispossession, Pueblo people still recognize the places where events of stories took place. "Stories are most frequently recalled," writes Silko, "as people are passing by a specific feature or the exact location where a story took place." Remembering, telling, and retelling events that occurred near "a rock or tree or plant found only at that place," Silko describes a process of imagination that binds Pueblo people to "some universal aspect of location" ("Interior and Exterior

Landscapes" 33). Although Silko's writing refers specifically to Southwest
United States' Laguna Pueblo culture, Ursula Heise notes that tribal people
often adhere to the paradigm in which identity, story, and landscape are
inextricably connected (Heise 33).

According to Vandana Shiva, while some scholars emphasize the importance of focusing on local modes of belonging, others hold out "global"
identifications as a possible alternative in many environmental and developmental discourses. Shiva sees this debate as a false impasse because, in
reality, the so-called promise of globalization masks a particular desire on the
part of multinationals to exploit local cultures, communities, and resources
(Shiva and Mies, "Introduction" 9). Nevertheless, as Leslie Silko notes, for
the overwhelming majority of Native Americans who live in a Diaspora,
along with those people scattered by the European slave trade, not all experiences people associate with place are so uncomplicated ("Auntie Kie" 86).

Without denying that affirmations of local ties play an important role in
environmental struggles, Silko, along with other Native women writers, of
necessity comes to terms with what Heise terms "deterritorialization," a process that refers to detachment of social and cultural practices from their ties
to place (51). For Southwest American Indian tribes, creation/emergence and
migration stories forever bind the people to a specific place. "From the time
we are very young, we hear these stories," explains Silko, "so that when we
go out in the world, when one asks who we are or where we came from, we
immediately know" ("Language and Literature" 51). This "web of memories," really "ideas that create an identity" ("Interior and Exterior Landscapes" 51), become so much a part of the self that Silko's move to Tucson
provided a renewed sense of her relationship with the land. In her memoir
Turquoise Ledge, she recounts her growing familiarity with the rattlesnakes,
wild bees, and other animate/inanimate beings that surround her house (114).

In *Yellow Woman and a Beauty of the Spirit* (1996), she elaborates on that
voluntary move as well as the vast Diaspora resulting from various peoples'
involuntary removal from their homelands. Rejecting her previous notion
that rigid boundaries designate a homeland, Silko begins to understand that
"invisible lines" do not define ownership. Her imagination sparked by the
growing hospitality of snakes around her home, Silko recalls that in the old
days her people also moved over the wide spaces of the desert region. Just as
they understood that there are no boundaries between the people and the
land, she recognizes now what she terms a profound understanding of
"home" ("Auntie Kie" 86). Not meant to negate her relatives who were
relocated from their homelands so that they might live in impoverished areas
of cities or rural areas, Silko's acknowledgment that "all places and all beings of earth are sacred" provides an alternative framework for the contemporary reality of globalization. "No part of earth is expendable," explains
Silko, and so "the earth is a whole that can't be fragmented." It follows, then,

that the universal rights of Mother Earth trumps what Silko calls the "destroyer's mentality" that seeks to exploit all natural resources as well as human labor ("Tribal Councils" 94).

Leslie Silko suggests ways of imagining localism from a larger environmental perspective. Pointing out the danger of privileging some places as sacred when all are blessed ("Tribal Councils" 94), she develops a perspective akin to what Ursula Heise defines as environmental advocacy that "encompass[es] the planet as a whole" (9). Luci Tapahonso also reflects on a "sense of place" that engages with rapidly engaging realities of tribal life. Her approach ponders eco-cosmopolitanism from the perspective of colonial, marginalized peoples who have migrated to urban areas but continue to follow indigenous traditions and local knowledge. Although Tapahonso left the Navajo Nation to accept a university position in Kansas, she carries with her the four sacred peaks that define the people. "The Holy People set the intricate and complex patterns of Diné life when they decreed that we should love her," she explains, "but they took care to ensure that those concepts could be integrated into modern life." Far away from the four mountains, Tapahonso continues to clean her house in the "proper way," starting from East to West, a pattern she repeats while stirring a pot, participating in discussion, or preparing for a First Laugh Dinner. Even in Kansas, she says, when a rainbow appears, she knows that the Holy People have returned in order to hold all beings in the "radiant curve of their care and wisdom" wherever her people might be ("A Radiant Curve" 7).

In her collection *The Women are Singing* (1993), Tapahonso takes a literary approach to the problem of maintaining cultural practices in a changing world. Just as the waterways she followed on her journey evolved from the Missouri River, so, she explains, the Navajo Nation is the source of her writing wherever she might live (xi). Looking back to Silko's assertion that remembering the stories allow the people to go out into the world with the security of knowing who they are, Tapahonso, too, affirms that her work owes its source to stories told to her as a child. For many people, she concludes, residing away from home, writing becomes the means "for restoring our spirits to the state of 'hozo,' or beauty," which forms the core of Navajo life. "As Navajo culture changes," she writes, "we adapt accordingly" (xii).

While some Native women writers foreground increased mobility as the main ingredient of modern life, Louise Erdrich's fourteenth novel, *The Round House* (2012), highlights ways that the experience of place has been transformed especially for those individuals and communities that stay put. It comprises several issues covered in this chapter: the primacy of family ties and unity of nature, both seen as inherent in resisting dominant global forces, and the appropriation of the colonizer's language and literary forms as a means to achieve self-determination. Moreover, *The Round House* is a multiethnic novel that deserves to be read as environmental literature partly be-

cause its complexity explicitly counteracts any tendency to romanticize this
Ojibwe community's sense of place. Set on the North Dakota reservation that
informs her previous books, it tells the story of Joe, a thirteen-year-old who
seeks justice after his mother, Geraldine, is brutally attacked. Focusing on the
harsh realities of contemporary life where Ojibwe and whites live uncomfort-
ably together, it also embraces many of the issues woven throughout this
article.

Mobility, particularly the involuntary removal or later relocation of native
people, informs the writing of many indigenous women writers. But the
experience of Erdrich's characters is that of staying in one place while experiencing the changes that global modernity brings to them. For example,
Geraldine's spiritual rejuvenation takes place in the garden, that "middle
place," says Joni Adamson, in which nature and culture are not separate, but
bound together (48). But the bedding plants that Joe's father believes will
also bring Geraldine back to life are bought at a "tumbledown hothouse"
miles away from the reservation. The lilacs, too, that Bazil Counts hopes will
inspire his wife to leave the seclusion of her room are not indigenous but
were planted by a reservation farm agent many years before (86).

Although the Ojibwe's sense of place has been altered first in the context
of Federal policy then later by globalization, global networks and exchange
have not altered their connectedness to the land. "I began to like what I was
doing," Joe says, as he feels the dirt that remains "most deep down below the
surface." Choices of products to cultivate and food to eat might now be
dictated by needs of the first world, but Joe still looks to the earth that has
long sustained his people to "drain [his] rage" over his mother's brutal rape
(91).

Just as the garden is shaped by processes and products that originate
elsewhere, the community, too, is not ethnically pure but mixed. Both
counteract any tendency to romanticize (environmentally or culturally speaking) the Ojibwe people who live there. For example, the villains in this work
are the Larks, Anglo intruders who live uneasily on the edge of reservation
life. Yet their daughter, Linda, grows up "adopted in" by a tribal member
who is unwilling to give up this damaged twin after she is abandoned by her
parents (50). As an adult, Linda, who continues to live on the allotment,
becomes a crucial link in the fight for justice against Geraldine's attacker
(51).

For Erdrich, then, sustainable communities are built from contemporary
engagement in particular places reinforced by past traditions. Very much like
the Yaqui, removed to a Tucson neighborhood from their original homelands, but still conscious, says Silko, of their identity as Yaquis ("Auntie
Kie" 90), so this community knows its place. "Indians know other Indians
without the need for a federal pedigree," Joe explains, "and this knowledge—
like love, sex, or having or not having a baby—has nothing to do with

government" (31). Despite encroachment from the dominant society, this community shares what Silko calls a "shared consciousness of being a part of a living community that goes on and on" ("Auntie Kie" 90).

While urban, middle-class women, according to Vandana Shiva, do not embrace this self-in-relation with either nature or colonized peoples (Shiva and Mies, "Introduction" 5). Poet-writer-musician Joy Harjo envisions her kitchen table as a place where food, creativity, and extended family write (19). This alliance joins the gifts of nature with a variety of cultures. What women—both North and South—have in common is an earthiness divided by thousands of miles yet also united by work that binds women to each other and to nature. "To understand the direction of a society," says Harjo, "one must look toward the women who are birthing and intimately raising the next generations" (21). Reflecting what Shiva calls profound human needs—for food, shelter, knowledge, freedom, leisure, and joy (Shiva and Mies, "Introduction" 6)—those shared concerns point the way towards a more universal social change agenda. By struggling for common ground based on resistance against dominant global forces, which permits the North to dominate the Global South, men to dominate women, Native women writers thus foster transformation of values based on the control of both people and resources worldwide for the sake of monetary profit.

Respect for the integrity of all life forms is at the core of eco-feminism, a holistic worldview that sees nature as a web of life. An ecofeminist perspective informs Leslie Silko's *Yellow Woman and a Beauty of the Spirit*, a collection that is organized, she says, like a spider's web. Rejecting the Enlightenment notion that liberation and happiness depend on increasing emancipation of nature, Silko "think[s] of the land, the earth, as the center of the spider's web" (21). It follows that cultural identity, storytelling and imagination all radiate from the center of her web/writing. In this way she highlights contradictions between what Shiva terms "the Enlightenment logic of emancipation and the eco-logic of preserving and nurturing cycles of regeneration" (6).

Ecofeminism also asserts the connection of theory and practice, academics and activists, all necessary to assure an alternative to existing capitalist systems. For Native American women writers, in particular, any focus on environmental struggle incorporates the desire for acceptable wages, education, and health care. For example, Erdrich's Round House refers to a sacred space and place of worship for Ojibwe people, but it is here that Geraldine's rape takes place. Three classes of land meet there—tribal trust, State, and fee—signifying a tangle of laws that hinder prosecution of rape cases on many reservations today (160). It also represents what is at the core of feminism, that capitalism exploits not only the land but also those considered associated with it, women and Others, in particular.

Aware of the connection between patriarchal violence against women, other people and nature, Erdrich's characters understand from their own historical experience that such threats to life itself might require violence as defense against violence. When Bazil suspects his son's killing of Geraldine's attacker, he says that it represents a "wrong which serves an ideal justice." Not only did it settle a legal problem, that "unfair maize of land title law by which Lark could not be prosecuted" (306), but it also allowed his mother to work in her garden without having to look over her shoulder in fear of the return of her attacker (252). Joe's action allows his mother's spirit to return to her in the garden, a place not to be confused with ethereal spirituality, but where sustenance requires not only sweat but also struggle to displace residual structures of colonialism and oppression.

According to Ursula Heise, First World environmentalism focuses on a set of values, including preservation of natural resources that are achievable only after arriving at a certain level of wealth. In the Global South, by contrast, less affluent communities struggle to gain control over natural resources needed for necessities of everyday life (59). Some theorists, however, such as Vandana Shiva, along with certain Native women writers, have incorporated spirituality as well as Indigeneity into this materialist struggle aimed at the preservation of local ecological systems. In *The Turquoise Ledge*, Leslie Silko's sense of inhabitation of place, of intimate mutual interaction and responsibility for both animate and inanimate facets of her environment, stands in stark contrast to the dominant culture of development. Because she feels profound obligations to her surroundings, Silko takes action when one of her neighbors decides to appropriate local boulders and rocks to be used only for landscape purposes in his yard. Perceiving humans as merely one aspect of the environment, Silko bemoans that "the man and the machine" (302) and the violence he perpetrated against the arroyo had not only disrupted the area but also the animals who called it home. In contrast to Joe's choice of violence in defense of violence perpetrated against his mother, Silko resorts to painting small white crosses, symbols of the Star Beings, on the remaining rocks and boulders (308). One year later, she observed that the "boulders with the crosses, even the ones the men flipped over, were beginning to lose the appearance of sudden violence" (318). In this way, Silko's local action serves as a symbol for continued large-scale resistance against resource extraction of natural resources on tribal lands.

Aware that those who control the land have also controlled its story, women writers included here understand the process of writing as resistance within the larger framework of community. Complicating this process, Arundhati Roy notes that words have been "ursurp[ed]" (6), thus turning language upside down to mean the opposite of its original intent. For example, when "market" no longer means a place to buy goods, those who oppose are labeled "anti-progress." This appropriation of language, Roy concludes, this cultural colonialism that consists of "usurping words and developing them like weapons," has been one of the most "brilliant victories" of the "tsars of the new dispensation" (5). In the face of this continued dispossession, Native American women writers understand that in order to become empowered rather than victimized by displacement, they must gain power to change the history of conquest and colonization by appropriating the language of the colonizer.

"Reinventing the enemy's language," then, as in the title of Joy Harjo and Gloria Bird's anthology of Native American women's writing, becomes advice to transform those ways of being that were stolen. Just as Arundhati Roy asserts that words have been transformed to mean the opposite of what they mean, so Harjo notes that the "colonizer's languages" often appropriated tribal words or demeaned them. Now the work involves taking back the symbols of tribal cultures—"our own designs and beadwork, quills if you will"—in order to "heal, to regenerate, and to create" an alternative vision, one that replaces the marginalization of people and degradation of the environment with another story. "To speak at whatever the cost," Harjo concludes, "is to become empowered rather than victimized by destruction" (22). Tribal cultures have always known the healing potential of language, so what better source for eco-critics interested in forging links between various groups of women who share a desire to sustain their ways of life.

In *The Round House*, Louise Erdrich illustrates that struggling for alternative visions of the world requires that tribal people must understand how the official world works in order to change it. When Bazil informs his son that his own father's law book contains the tools to resist social and environmental injustices, he is instructing Joe to appropriate for his own use the very weapons that dispossessed his people (228). Engaged in a seeming contradiction—working to preserve a traditional way of life with the aid of ancient stories and modern texts—Bazil tells his son that "we are trying to build a solid base here for our sovereignty" (229). In Bazil's view, there are no divisions between categories such as traditional and modern, nature and culture, local and national. Moving between the two landscapes, not only the official but also the vernacular, Bazil invites his son into an in-between space by telling him stories of the past at the same time that he is showing him how he might use Western law to enlarge jurisdiction on the reservation.

The reconstruction and restoration of identity—dependent upon a rediscovered sense of place in relation to the community and the land—are at the heart of Erdrich's writing. Itself a fluid, always shifting space, the Round House transcends geographic location by exerting its own will, much like Native people who resist colonization by refusing to be contained, clearly mapped, or controlled by the dominant culture. When Joe says to his father, "Why do you do it . . . Why bother?" (229), Bazil replies by turning a moldy uneaten casserole into a history lesson for his son. As Bazil balances eating utensils precariously one by one on the frozen dish, he develops a compelling critique of Euro-American law while insisting on using the tools of that culture to further his own goals. As the casserole begins to "ooze and thaw" (229), he probes, blurs, and replaces rigid definitions by asserting his own space. "We try to press against the boundaries of what we are allowed, walk a step past the edge," Bazil explains, in order to "build a solid space here for our sovereignty" (229). Here Bazil appropriates and inverts the European notion of marked, fixed and terminal territory, turning it into a space between opposing worlds, a kind of reconciliation between Native American and the mainstream that Bazil wants his son to learn. In *The Round House*, then, Erdrich appropriates her own territory, offering a space for transcending and reconstructing history.

Writing about jurisdictional issues on American Indian reservations, Erdrich examines how racism inflicts injuries both on the bodies of Native women and the body of nature. Appropriating rather than dismissing the dominant structure, she shows how Bazil might offer this insight to his son. Once this lesson is understood, Erdrich implies, Native people can resist the fractures of colonialism by defining themselves from their own perspectives, speaking for themselves, and reclaiming their right to decide the terms with which they will live in the world. In the end, Joe moves away from his father's law books to return to the stories that his grandfather told him in his sleep (307). In this way Erdrich weaves one generation into the fabric of the next, ensuring thus the continuity of healing traditions that are threatened today by globalization.

WORKS CITED

Adamson, Joni. *American Indian Literature, Environmental Justice, and Ecocriticism.* Tucson: U of Arizona P, 2001. Print.

Allen, Paula Gunn. "The Woman I Love Is a Planet; the Planet I Love is a Tree." *The Sweet Breathing of Plants: Women Writing on the Green World.* Ed. Linda Hogan and Brenda Peterson. New York: North Point, 2001. 79–86. Print.

Erdrich, Louise. *The Round House.* New York: Harper Collins, 2012. Print.

Harjo, Joy, and Gloria Bird, eds. *Reinventing the Enemy's Language: Contemporary Native American Women's Writings of North America.* New York: Norton, 1998. Print.

Heise, Ursula. *Sense of Place and Sense of Planet: The Environmental Imagination and the Global.* New York: Oxford UP, 2008. Print.

Hogan, Linda. *The Book of Medicines.* Minneapolis: Coffee House, 1993. Print.

———. "The Two Lives." *I Tell You Now: Autobiographical Essays by Native American Writers.* Ed. Brian Swann and Arnold Krupat. Lincoln: U of Nebraska P, 1987. 231–51. Print.

———. *The Woman Who Watches Over the World: A Native Memoir.* New York: Norton, 2001. Print.

Hogan, Linda, and Brenda Peterson, eds. *The Sweet Breathing of Plants: Women Writing on the Green World.* New York: North Point, 2001. Print.

Roy, Arundhati. *Field Notes on Democracy: Listening to Grasshoppers.* Chicago: Haymarket, 2009. Print.

Shiva, Vandana, and Maria Mies. "Introduction: Why We Wrote the Book Together." *Ecofeminism*. London: Zed, 1993. Print.

Silko, Leslie Marmon. "Auntie Kie Talks About U.S. Presidents and U.S. Indian Policy." *Yellow Woman and a Beauty of the Spirit*. New York: Simon and Schuster, 1996. 80–85. Print.

———. *"Interior and Exterior Landscapes: The Pueblo Migration Stories." Yellow Woman and a Beauty of the Spirit*. New York: Simon and Schuster, 1996. 25–48. Print.

———. "Language and Literature from a Pueblo Indian Perspective." *Yellow Woman and a Beauty of the Spirit*. New York: Simon and Schuster, 1996. 48–60. Print.

———. "Introduction." *Yellow Woman and a Beauty of the Spirit*. New York: Simon and Schuster, 1996. 25–48. Print.

———. "Tribal Councils: Puppets of the U.S. Government." *Yellow Woman and a Beauty of Spirit*. New York: Simon and Schuster, 1996. 92–96. Print.

———. *The Turquoise Ledge*. New York: Penguin, 2011. Print.

Tapahonso, Luci. *A Radiant Curve*. Tucson: U of Arizona P, 2008. Print.

———. "A Radiant Curve." *A Radiant Curve*. Tucson: U of Arizona P, 2008. 7–27. Print.

———. *The Women are Singing*. Tucson: U of Arizona P, 1993. Print.

Chapter Eight

Redefining Modernity in Latin American Fiction: Toward Ecological Consciousness in *La loca de Gandoca* and *Lo que soñó Sebastian*

Adrian Taylor Kane

Octavio Paz has stated, and literary critics have amply documented, that the concept of modernity has been a prominent concern of Latin American intellectuals since the nineteenth century (5). A cursory glance at the terminology used by authors and critics to denominate Latin America's various movements throughout the last 125 years—modernismo, posmodernismo, vanguardia, novela moderna, and ficción posmoderna—provides a sense of the persistent desire of Latin American authors to engage the topic of their nations' positions within modernity. Each of these moments in Latin American literary history reflect what Raymond L. Williams identifies as a "desire to be modern" that manifests itself in unique forms according to historical context and aesthetic aspirations (369). Over the past two decades in Latin America, a corpus of work has emerged that suggests that it is now possible to speak of the environmental novel as a bona fide subgenre of Latin American fiction that undoubtedly merits careful analysis itself.[1] In what follows, I would like to suggest that the development of the Latin American environmental novel is a further manifestation of Latin American intellectuals' historical desire to conceive new paradigms of modernity and, moreover, that the emergence of such novels is inextricably linked to Latin America's position within the economic and geopolitical construct of the Global South. That is, Latin American environmental fiction arises in the context of a globalized, neo-colonial order as an aesthetic forum in which authors elaborate counter-discourses to the economic models of modernity that have often been violently imposed by the Global North.

Although this chapter is necessarily limited in scope, I offer two novels as examples of my thesis: Guatemalan novelist Rodrigo Rey Rosa's 1996 *Lo que soñó Sebastián* (What Sebastian Dreamt) and Costa Rican writer Anacristina Rossi's 1992 *La loca de Gandoca* (The Madwoman from Gandoca). *Lo que soñó Sebastián*, framed within the context of a globalized black market, presents a conflict between an environmentalist, who moves from the city to the Guatemalan jungle, and his neighbors, who are engaged in illegal hunting. By reading this novel through the theoretical framework of deep ecology, I will demonstrate how it postulates Guatemala's new utopian project as a shift from anthropocentric to ecocentric thinking that would render the country's biotic diversity and unique cultural heritage its most valuable patrimony. Rey Rosa's novel is underpinned by an understanding of the natural world that coincides with many of the fundamental ideas of deep ecology, and, in particular, with its emphasis on the analysis of environmental threats in relation to complex systems—whether biological, economic, or political (5). My analysis of *Lo que soñó Sebastián* specifically highlights the importance of reading environmental literature from the Global South with a heightened awareness both of the role of international markets in local environmental issues as well as the devastation that is often wrought by the economic power of the nations of the Global North. The North/South dynamic is also central to *La loca de Gandoca*, in which the narrator interweaves the story of her legal battle against the development of a wildlife refuge on the Caribbean coast of Costa Rica with that of her relationship with her husband and his struggle with alcoholism. By drawing on recent theories of addiction and focusing my analysis on the relation between the secondary storyline of her husband's illness and the primary narrative of the explicit environmentalist critique, I argue that Rossi's novel portrays Costa Rica's neoliberal economic model as an addiction to foreign capital that is destroying the country's natural resources like a disease. Both novels present counter-discourses to external economic forces that threaten their nations' natural resources and, in doing so, engage in the redefinition of modernity by proposing the possibility of creating more ecologically sustainable economies.

Rodrigo Rey Rosa's novel *Lo que soñó Sebastián* is the story of a man who leaves his city life behind to begin anew in a remote region of Guatemala's Petén jungle. In the opening passage the protagonist Sebastián describes his newly constructed house as a masterpiece of carpentry, made solely of wood without any nails. Its design, he continues, is such that when inside, he feels as if he were outside. From the very beginning of the novel, therefore, Sebastián's longing for a connection to the nonhuman natural world is evident. The theme of the relationship between nature and modernity is present throughout much of the novel, as evidenced by the constant pressure that the global economy places on local resources. It becomes particularly salient, however, when Sebastián recalls his father's words during a conversation about his move to the jungle: "Only a savage that had nothing to lose could live in such conditions. Or a man as modern as you" (45).[2] Because this quote is isolated as a brief analepsis, the reader is not privy to any further development of his father's ideas. As in many passages from Rey Rosa's novels, a certain level of ambiguity places the onus on the reader to decipher or construct the text's meaning. By the novel's conclusion, however, it becomes clear that what is implied in this line is a radical redefinition of modernity that carries with it a demand for behavioral changes in an effort to move toward sustainable living.

Sebastián's environmental orientation is evident in the early pages of the novel when he reminds his friend Juventino that hunting on his property is strictly forbidden. This fact underpins the novel's central conflict between Sebastián and his neighbors, the Cajals, who make a living by illegally hunting several types of animals. During the course of Sebastián's conversation with his friend, Juventino spots a caiman, and ignoring Sebastián's orders, prepares his rifle in pursuit of the reptile. Juventino disappears momentarily, and Sebastián hears a shot, two men voicing insults, and then another shot. After the men flee, Sebastián moves in the direction of the gunfire, and discovers not only the dead caiman, but also Juventino's corpse with a bullet lodged squarely between his eyes. There is little doubt that the perpetrators are the Cajals, as one of their dogs remains behind at the scene of the crime. This violent event quickly puts an end to any romantic illusions that Sebastián may have had about an idyllic return to nature, and underscores the ruthless and reckless character of illegal hunters. Illegal hunting is one of several of Guatemala's current environmental and cultural issues presented in *Lo que soñó Sebastián*, and as the novel unfolds it becomes evident that even remote regions of dense jungle cannot escape the ubiquitous reach of the global market.

A primary objective of many environmental writers is to move readers from an anthropocentric view of creation—in which "the interests of humans are of higher priority than nonhumans"—to an ecocentric view in which "the interest of the ecosphere must override that of the individual species" (Buell 134, 137). Indeed, this is also one of the tenets of deep ecology, which emphasizes that all organisms are part of a web of biospheric relations and rejects the image of the human being in the environment in favor of the notion that humans and all other organisms are inseparable from the other elements the biotic system (Naess 5). In *Lo que soñó Sebastián*, two brief passages that come after Juventino's murder remind the reader that despite the common tendency to create a false and often hierarchical divide between human beings and the rest of the natural world, humans are as much a part of nature as any other animal. The first passage reads as follows: Sebastián "cautiously approached the small clearing in the palm forest where Juventino lay a few steps away from the caiman, each one with a dark, circular red wound in the head. The dog was less interested in the man than the great reptile" (14). That the dog is more interested in the dead caiman than the human being suggests that despite Juventino's death being a great human tragedy, to nature this is no different than if any other being had died. When Sebastián returns to the scene later, the narrator describes what he sees: "The bullet that had killed Juventino had entered between his two eyebrows. His face, the color of wax, was covered with black ants" (16). This image of ants feasting on Juventino's cadaver reinforces the fact that humans are one among many members of the natural world, and thus coincides with deep ecologists' call for a move from anthropocentrism to ecocentrism. Moreover, it recalls Horacio Quiroga's 1920 classic crillolista story "El hombre muerto" (The Dead Man) in which, upon the protagonist's death, his horse subsequently walks by his corpse as if nothing had happened. For Jonathan Tittler, Quiroga's story implies that "what bears all the signs of tragedy for the man . . . is no more than a routine day for the cosmos, which continues its cycles of life and death, creation and destruction, with absolute impassivity" (Tittler 16). The images of the disinterested dog and the ant attack on Juventino's corpse in *Lo que soñó Sebastián* produce a similar effect by moving human beings from a privileged space at the fore back into the biospherical matrix of what Arne Naess, a pioneer of the deep-ecology movement, describes as "the relational, total field image" (3).

The topic of illegal hunting continues when Sebastián returns to the crime scene the next day to ensure that no one has removed the dead caiman from his property. When he discovers that it is gone, he follows tracks to the Cajals' property where he finds the skinned animal hanging out to cure with a piece of deerskin and the pelt of a margay (a small spotted wild cat whose scientific name is leopardus wiedii). Sebastián, engaging the Cajals' brother-in-law in conversation, asks him if it isn't illegal to hunt these animals. "There are laws against this, yes, but they haven't arrived here," responds Benigno (48). The implication, of course, is that the fact that it is illegal to hunt these animals is of little importance to the Cajals. Unsurprisingly, the principal motive for the Cajal family to ignore hunting laws is profit. The narrator reveals that two men, one a police captain and the other a military commissioner, who help Roberto Cajal frame Sebastián for the murder of Juventino, are part of a local culture of corruption that financially supports the illegal hunters. The narrator states: "The soldier was addicted to tapir meat, impossible to find in the market; and the judge's son had a famous restaurant in Flores specializing in game, whose principal provider was Francisco Cajal. (The mayor of Sayaxché, for his part, was the host to a norteamericano that came each year to buy skins)" (56). The Cajals thus have little motivation to obey the law, as the very people that are entrusted to enforce it are accomplices in their illegal activity. Furthermore, it is revealing that the demand for the illegal skins comes from the Global North. The term norteamericano, used by Latin Americans (including Mexicans) to refer U.S. citizens, indicates that the North/South divide is not just a construct imposed from outside, but that Latin America has internalized a sense of otherness that is defined culturally, rather than by a country's location on the map. As is the case in this passage, economic power is often an implied element of the otherness that underpins this term. By introducing this faceless American into the narrative, the implied author represents one of the many ways in which the Global North is entangled in the depletion of local resources in Latin America.

Specifically, *Lo que soñó Sebastián* includes three animals in the plot whose major existential threats include illegal hunting—caimans, margays, and tapirs. By weaving these animals into the storyline as characters of equal importance as humans and portraying their unjust treatment, the implied author acknowledges what deep ecologists refer to as "biospheric egalitarianism," that is, the right of all organisms to live and flourish (Naess 5). The tapir, a large, hoofed, hog-like animal with a flexible snout, is not as central to the plot as either the caiman or the margay. Nevertheless, the soldier's love of tapir meat, which is illegal to hunt in Guatemala, demonstrates the Cajals' links to the black market and is evidence of their illicit activity and disregard for the local ecosystem. Perhaps it is because of its depleted population that no live tapirs appear in the novel.

The animal killed at the beginning of the novel is a caimán, which could refer to a variety of crocodilians, but one particular crocodilian that is reported to have a depleted population in Petén, where the novel takes place, is Morelet's crocodile, which is legally protected under the 1970 General Law of Hunting in Guatemala (Thorbjarnarson et al. 64). Due to the high quality of its skin, this particular species was severely depleted by hunting during the middle of the twentieth century, and "the trend continues through further illegal and indiscriminate taking of their skins" (Britton). In *Lo que soñó Sebastián*, the fact that the mayor of Sayaxché hosts a U.S. merchant who makes an annual trip to Guatemala to purchase skins of such animals, is an excellent example of how the global market reaches even the deepest corners of the Petén jungle.

The margay is another animal that is central to the plot of Rey Rosa's novel. One day Roberto Cajal appears on the property of Sebastián's neighbors the Howards, who own an inn near an archaeological site of Mayan ruins. He is carrying a bag containing three margay kittens that he captured after killing their mother. The inn-owner is furious, but concedes that it would be fruitless to report him to the authorities. Véronique, a guest at the inn who has travelled to the jungle to visit Sebastián, purchases one of the kittens from Roberto with the intention of returning to Paris with it. When she appears at Sebastián's residence with the animal, he is clearly unhappy, and becomes even more so when he learns that she has purchased it from his enemy. For Roberto the margay is a symbol of the survival of the fittest and a way of justifying his existence as a hunter. "It's destiny," he muses, "The place of each of us in the world. I'm like that little cat. The more and better it can kill, the better it will live. [. . .] As long as there are animals in the jungle, I will be a hunter" (84). The comparison with regard to hunting behavior is logical. Indeed, the analogy seems to be reinforced in the text when the narrator describes the cat's instinctive hunting behavior, as it repeatedly climbs onto Sebastián's ceiling beams and pounces on his head. The flaw in Roberto's thought, however, is revealed in the excess of his statement that as long as there are animals in the jungle, he will continue to hunt. What he fails to recognize is that the very existence of the species that he intends to continue hunting depends in part on his own actions. Roberto's vision is undoubtedly anthropocentric, and, more accurately, egocentric. In contrast to Roberto's view, in an ecocentric outlook, according to Naess's tenets of deep ecology, "the so-called struggle of life, and survival of the fittest, should be interpreted in the sense of ability to coexist and cooperate in complex relationships, rather than ability to kill, exploit, and suppress. 'Live and let live' is a more powerful ecological principle than 'Either you or me'" (4). The novel's portrayal of Roberto as an egotistical outlaw in contrast to Sebastián's altruism suggests that the struggle for survival should be understood within the context of the complex web of relations that deep ecology emphasizes and Sebastián clearly understands. In other words, Roberto's anthropocentric way of thinking must give way to a more ecocentric perspective.

In Roberto's view, Sebastián's battle with the hunters, who Roberto claims "love the animals" more than Sebastián, is misguided. Roberto suggests that Sebastián should focus his efforts on other ecological problems such as the farmers, who "level the trees in the forest to make pastures" (84). Roberto's claim that he cares more about the animals than Sebastián is dubious, given his refusal to modify his hunting behavior according to laws based on scientific knowledge. Nevertheless, his argument brings the significant environmental issue of Guatemalan deforestation to the fore.

Daniel Faber, who has analyzed Central American environmental issues in relation to the region's history of imperialism, submits that nineteenth-century liberal agrarian reforms to facilitate expansion of the coffee market "triggered massive migrations by displaced Indian villagers," which resulted in the destruction of "huge tracts of previously undisturbed oak-pine forests above the upper altitude limits for coffee production" as well as "other forested (but more ecologically fragile and marginal) agricultural lands, which they cleared for their own small subsistence plots" (27). A similar dynamic occurred in Central America throughout the twentieth century, as peasants were displaced to make way for the expansion of banana plantations and beef-producing cattle pastures to meet the ever-increasing demand for North American markets. To this day, displaced peasants seeking land for ranching and farming continue to burn sections of rainforest (Dawn of the Maya). According to ecologist Kristen Silvius, "pasture land in the Guatemalan Petén increased by more than 200,000 hectares between 1979 and 1996, largely at the expense of previously undisturbed rain forest" (100). As of 1998, biologists measured the rate of forest destruction in the Maya Forest at more than 80,000 hectares per year (Primack et al. xvi).

This process is represented in *Lo que soñó Sebastián* in a passage that describes a pasture that has been cleared in the midst of the jungle: "The jungle ended abruptly in a barbed-wire fence; on the other side, in a pasture where a ceiba tree was growing that seemed not to project any shade, grazed a herd of white and beige zebus" (88). The barbed wire fence in this image serves as a metaphor for the human alteration of the environment and distinguishes the natural landscape of the jungle from the semi-natural landscape of the pasture. That one of the characters is bitten by a venomous snake in the passage that follows this description is perhaps a symbolic suggestion that such human destruction is indeed a risky endeavor that may very well bring about unintended ecological consequences. Behind this seemingly insignificant description of a pasture in the jungle lies a history of imperialism that has led to centuries of continual displacement of subsistence farmers and rampant deforestation. Once again in line with the philosophy of deep ecology, the novel begs the reader to consider Guatemala's environmental issues in the context of the complexity of larger systems—whether biological, economic, or political (Naess 5).

Conservationists have noted that, "the biological and cultural richness of the Maya Forest is threatened by a variety of forces, including but not limited to the conversion of forests to agriculture; the plundering of archaeological sites; legal and illegal trade in wildlife, wildlife products, and cultural artifacts; logging; and extraction of other renewable and nonrenewable resources, including oil and mineral wealth" (Norris, Wilber, and Morales Marín 328–29). By incorporating several of these environmental and cultural issues into *Lo que soñó Sebastián*, Rey Rosas's novel sheds light on some of the most important social problems that Guatemala faces in the age of globalization. In the novel's denouement, Roberto Cajal is released from prison after serving a sentence for selling an ancient Mayan vase. When he returns home, he discovers that his family has changed its way of life. They have constructed corrals to raise tepeizcuintes (large rodents known also as jungle rats) and have arranged to sell their meat to a local merchant. His uncle Francisco Cajal explains, "Now they don't let us hunt like we used to," to which Roberto responds, "I don't know how to do anything else except hunt. To do something else would be like being dead. I don't plan on changing"

(114). In the novel's final sentence, Francisco acknowledges that he understands his nephew's point of view, and Roberto suddenly observes that his uncle has aged significantly. This detail, symbolic of Francisco's individual transformation, suggests that despite his previous reckless behavior he has grown into the archetypical wise elder. At the same time, it indicates the suffering that comes with relinquishing a personal passion that has been an important part of his family's and likely his ancestors' culture. The novel thus implies that Guatemala, like Francisco Cajal, must make some difficult choices to preserve the nation's biological diversity, what Puerto Rican novelist Enrique Laguerre refers to as a "nation's most valuable patrimony" (qtd. in Paravisini-Gebert 120). The conclusion thus suggests that a shift in thinking towards a greater environmental consciousness is precisely the redefinition of modernity that is implied in Sebastián's father's comment at the beginning of the novel.

A similar demand for a paradigm shift is present in Costa Rican author Anacristina Rossi's 1992 novel *La loca de Gandoca*. In contrast to the lesser-known *Lo que soñó Sebastián*, *La loca de Gandoca* has received considerable critical attention as an example of the heightened ecological consciousness in Latin American fiction during the 1990s and 2000s.[3] As critics have amply documented the ways in which Rossi's novel corresponds with the theoretical frameworks of ecofeminism and deep ecology, the following reading of *La loca de Gandoca* focuses on the diegetic presence of addiction and the way in which the novel establishes a parallel between addiction and environmental destruction. Specifically, I argue that the lexicon of addiction is a central rhetorical strategy in *La loca de Gandoca*'s critique of neoliberalism as Costa Rica's economic model of modernity.

La loca de Gandoca is the story of protagonist Daniela Zermatt's struggle to halt the illegal development of a tourist district in the Gandoca Wildlife Refuge on the Caribbean coast of Costa Rica. Despite the fact that the construction of the new community violates a series of constitutional laws and international treaties, the bureaucrats to whom Daniela turns in her effort to protect the refuge insist that the matter is beyond their jurisdiction and continually redirect her from one governmental agency to another to no avail. The story of Daniela's tumultuous relationship with her alcoholic husband Carlos Manuel is interwoven into the main plotline of her environmental battle in defense of the refuge.

While it would be easy to dismiss the story of Daniela and Carlos's love affair as a textual element of secondary importance, it in fact fortifies the novel's central cultural critique by drawing a parallel between Carlos's battle against alcoholism and Daniela's struggle for environmental justice. In *La loca de Gandoca* the notion of alcoholism as a self-destructive disease becomes a powerful metaphor for the wave of economic development projects that is sweeping the country. As Kearns argues, "An evil has infiltrated each one that will eliminate them: in Carlos Manuel, her husband, it is alcoholism, and in Gandoca it is the entry of the tourist and housing development" ("Otra cara" 315). Indeed, early in the novel the text invites the reader to consider the parallel between the two storylines that Kearns astutely alludes to. Daniela reflects, "It was a strange coincidence that the tolling of the bells for the destruction of those latitudes and for your illness should occur at the same time. That is why I didn't realize that the bells of the Refuge were sounding an alarm. Desperate, I could only hear the words of your own destruction crashing against my skull" (33). This passage not only connects the two storylines as significant events in Daniela's life, it also suggests a similarity between them that is revealed through Rossi's choice of lexicon. Her use of destrucción (destruction) and enfermedad (illness) in this passage marks the beginning of the process of lexically linking the two stories in order to elicit the novel's central metaphor. The implication of the addiction metaphor is that the Costa Rican government is creating an imbalance or illness in its ecosystems through its own self-destructive behavior.

What the novel therefore suggests is that the desire for accumulation of capital through development is a force that is equally as powerful as alcoholic addiction. Throughout the novel Daniela emphasizes the greed that motivates the foreign investors:

> Everything was going to disappear to raise beauty salons, discotheques, stores, restaurants, tennis courts, pools, bungalows, level streets, and clean lots, without puddles, without those bothersome bushes that serve as food for the crabs, and, above all, without trees. Because the Italian company "Ecodollars" wanted to sell so many lots that each lot was tiny and a single tree was an impediment to construction. (70)

However, the addicts in the novel are not only the investors, but also the governmental officials that financially benefit from these transactions. As a result, those who are entrusted with the well-being of Costa Rica's citizens and the protection of their natural resources are seduced by the desire for profit, thus abandoning their ethical and professional responsibilities. The concept of the seductive power of this economic addiction to foreign capital is evident in the following passage:

> thousands of tons of trash, pesticides, merciless deforestation, illegal concessions to foreign companies, and hundreds of devastating mega-tourist projects like the massive hotel Serafin Cataló that had destroyed a mangrove, the beautiful Pánica River, a mountain, two indigenous cemeteries, and was appropriating an entire bay. And according to the authorities there was no way to regulate these investments because if foreign capital is regulated, it leaves. (113)

This quotation does well to capture the novel's central criticism of the government's attitude that environmental destruction, in and of itself, is not sufficient cause for detaining harmful development because it would deter foreign investors from conducting business in Costa Rica. In other words, Costa Rica's neo-liberal economic model has led to an unhealthy dependence on the influx of foreign capital that accompanies tourist development.

One consequence of this economic model is what Annelies Zoomers describes as "the foreignization of spaces" in the age of globalization, driven by a "contemporary global land grab" involving transactions often carried out by transnational corporations in Africa, Asia, and Latin America (429). One of the primary motives for this phenomenon, maintains Zoomers, is the construction of large tourist complexes (438). Because these transactions tend to be carried out by corporations controlled by the Global North, the result of this process is a North-South land grab. This process can be found to varying extents throughout Latin America, but Costa Rica has a been a prime target due to its relative political stability and its geographical accessibility from the United States. Over the last three decades, the construction of tourist complexes in Costa Rica has led to a monopolization of coastal property to such an extent that as early as 1990 it was estimated that 80 percent of beachfront property in Costa Rica was owned by foreign investors (Honey 134). As Rossi compellingly captures in *La loca de Gandoca*, the construction of tourist complexes frequently comes at a great cost to local ecosystems and has a significant impact on natural resources.

The mentality of the bureaucrats in the novel that continually advocate for tourist development is similar to the behavior of addicted individuals in the sense that the economic benefits of construction satisfy a short-term desire without considering the long-term consequences for local ecosystems. And while it is true that every day millions of humans make poor long-term decisions to satisfy a short-term desire and this does not make them addicts, what recent theories of addiction have posited is that addiction is a mental disorder in which voluntary behavior has become self-destructive (Heyman 20), thus affecting the individual's motivational system (West 174). Gene M. Heyman explains his theory in terms of local choices and global choices: "In local choice, choosing the better option means choosing the item that currently has the higher value. In global choice, the best choice is the collection or sequence of items that has higher value" (119). The drug or addictive activity is always the best option from a local perspective, but never from a global perspective (Heyman 156). Global choices are usually more difficult because they require a considerable effort and tend to be more burdensome (Heyman 157). When the repeated behavior of making local rather than global choices becomes a self-destructive pattern that is not easily broken, the individual runs the risk of becoming addicted. Heyman thus concludes that, "addiction is a disorder of choice" (173). Robert West similarly proposes that, "we have arrived at a theory of addiction in which the individual chooses in some sense to engage or not engage in the behavior" (73). He submits that in order to change the behavior of the addict it is necessary to change his or her motivations and, to achieve this, one must change his or her values or conscious beliefs about what is good or bad, correct or incorrect, useful or harmful (7, 189). Therefore, one of the primary objectives of educational interventions is to change the individual's values (189). With specific regard to financial addiction, Barbara Brandt maintains that, "Money-addicted individuals, organizations, and governing institutions regularly destroy real wealth—in the form of human health, the natural environment, or community well-being—especially over the long term, in order to make money in the short term" (81). This is precisely the scenario portrayed in *La loca de Gandoca*.

The notion of economic addiction in the novel's environmental storyline runs parallel to the story of Carlos Manuel's alcoholism throughout the novel, but it is the lexicon of destruction and illness that Rossi employs that intimately entangles the two plots in such a way as to lay bare the central metaphor of addiction. This is visible, for example, when, in a conversation with an architect and an economist, Daniela learns that the recommended density of development in eco-tourism should not exceed 15 percent, whereas the Italian company Ecodollars is building at a density of 95 percent. The economist explains, "they make a fortune selling those lots, they are selling destruction" (46). What this and other passages throughout the novel suggest is that if Costa Rica continues the down the path of unbridled tourist development, the very natural resources that attract the tourists and investors in the first place will be destroyed. In *La loca de Gandoca*, Daniela is one of the few who possess this long-term vision. She therefore understands that in both storylines an intervention to end the cycle of destruction is necessary. However, despite her repeated efforts to find help for her husband, he dies inebriated in a car accident. Once again drawing a parallel between her husband's disease and the destruction of the refuge, Daniela laments, "The trees are going to die and you are already dead" (70). As Rossi confirms, "there is definitely a parallelism between both destructions" ("RE: Una pregunta").

Rossi's portrayal of Costa Rica's addiction to foreign capital is different, however, from Carlos Manuel's addiction in the sense that he is conscious of his condition and attempts several times to stop consuming alcohol, whereas the government does not express even a minimal interest in altering the neoliberal economic model that is undermining the reason for the existence of the Gandoca refuge. When Daniela attempts to defend the environment, she is called a madwoman (44) and is accused of impeding the country's progress (95). One official asks her, "How are you going to stop something as extraordinary as an immense investment of millions of dollars for the single and silly reason that perhaps the law prohibits it?" (95) This way of thinking is another example of the way in which the government officials behave like addicts, as they are willing to overcome any obstacle that prohibits the satisfaction of their desire, including through unethical or illegal means. In other words, the government ignores the scientific evidence of the harmful effects of development in the refuge and refuses to adhere to its own policies that were established to prevent such destruction, all for the sake of short-term financial prosperity.

With respect to the metaphor of addiction to foreign capital, Daniela repeatedly intervenes in an attempt to change the government's values by trying to educate officials about the environmental impact of the massive tourist project. Nonetheless, she is unsuccessful in her effort to halt the development of the Gandoca Refuge. After eighteen months of bureaucratic battles, she decides that her last resort is to write about her experience: "The word is all I have left" (114). In the novel's final sentence, the reader realizes that *La loca de Gandoca* is the novel that Daniela writes.

In reality, the novel is based on the experiences of the author Anacristina Rossi, and the publication of *La loca de Gandoca* had a significant impact on the course of events in the Refuge.[4] According to Kearns, Rossi's novel has provoked a national dialogue in Costa Rica about the relation between environmental policy and corruption ("Otra cara" 334). What's more, according to Rossi, "Three years after the novel was published, in 1995, the Constitutional Court ruled in my favor and ordered the Ministry of the Environment to Protect the Gandoca Manzanillo Refuge and to halt immediately all development projects that had not yet materialized" ("Mad About Gandoca" 195). In light of the impact the novel has had on Costa Rican society, *La loca de Gandoca* is an excellent example of the potential of environmental literature from the Global South to present effective counters to hegemonic discourses of development. In contrast to Laura Barbas-Rhoden, however, who argues in *Ecological Imaginations in Latin American Fiction* that Latin American authors' critiques of the neo-liberal economic model indicate a "retreat from modernization," I would submit that the critique of this form of modernity that she astutely observes is not a retreat, but, rather, an attempt to redefine modernity to include a paradigm for sustainable living (*Ecological Imaginations* 14).

I have heard the question on more than one occasion about whether ecocritical approaches to Latin American literature are not just one more example of the North American academy imposing a foreign theoretical framework on works written in and for a different cultural context. Sufficient scholarship exists to reject this notion, and the two novels analyzed in this essay are further examples the organic development of an environmentally conscious Spanish American literature that arises in response to local threats.[5] As evidenced in the passages that I have cited from *Lo que soñó Sebastián*, Rey Rosa's novel displays several ideas that coincide with the fundamental tenets of deep ecology. Ultimately, his work posits Guatemala's new utopian project as a shift from anthropocentric to ecocentric thinking in which the country's biotic diversity and unique cultural heritage are its most valuable patrimony. Both *Lo que soñó Sebastián*, with its strong correlation to deep ecology, and *La loca de Gandoca*, in its counter-discourse to neoliberalism bolstered by the metaphor of addiction, highlight the reality that the environmental challenges of the Global South are inexorably entangled with the economic forces created by the Global North. As a form of resistance to the ecologically harmful dynamics of globalization, *Lo que soñó Sebastián* and *La loca de Gandoca* partake in a growing trend among ecologically conscious Latin American authors to redefine modernity. Authors such as Rey Rosa and Rossi are currently promoting a departure from previous forms of cultural modernization that often put the desire to be modern before the necessity for sustainable living. Their redefinition of the concept of modernity signals a significant transmutation of the desire to be modern in Latin American fiction.

NOTES

1. See Barbas-Rhoden's *Ecological Imaginations in Latin American Fiction* for insightful analysis of several environmental novels and Kane (234) for a brief overview of several more works from this period.

2. All translations in this essay are my own.

3. See Barbas-Rhoden, Kearns, and Postema.

4. According to Barbas-Rhoden, "Around the time of the publication of Rossi's novel, the Gandoca-Manzanillo Reserve in particular was at the heart of a controversy involving illegal development and protected lands. In an irony worthy of fiction, the development project was headed by the architect of the 1992 Earth Summit in Rio de Janeiro. According to journalist Martha Honey, the wealthy Canadian businessman Maurice Strong and his company Desarrollos Ecológicos were finishing a $35 million resort within the reserve at the same time that the Rio Summit was opening (44). The uproar in Costa Rica centered on Strong's lack of title to the land and the trampling of rights of the indigenous population, since some of the project fell within the Kéköldi reserve and had not been approved by the indigenous association" (Ecological Imaginations 129).

5. See, for example, Barbas-Rhoden, Binns, French, Heffes, Kane, Kearns, Rivera-Barnes and Hoeg, and White as well as the special issues of *Review: Literature and Arts of the Americas* 86 (Fall 2013) on "Eco-Literature and Arts in Latin America" and *Hispanic Journal* (Fall 1998) as primary examples.

WORKS CITED

Barbas-Rhoden, Laura. *Ecological Imaginations in Latin American Fiction.* Gainesville: UP of Florida, 2011. Print.

———. "Greening Central American Literature." *ISLE: Interdisciplinary Studies in Literature and Environment* 12.1 (2005): 1–17. Print.

Binns, Niall. *Callejón sin salida?: la crisis ecológica en la poesía hispanoamericana.* Zaragoza, Spain: Prensas Universitarias de Zaragoza, 2004. Print.

Brandt, Barbara. *Whole Life Economics: Revaluing Daily Life.* Philadelphia: New Society Publishers, 1995. Print.

Britton, Adam. "Crocodylus moreletii." Florida Museum of Natural History. Web. 28 May 2013.

Buell, Lawrence. *The Future of Environmental Criticism: Environmental Crisis and Literary Imagination.* Malden, MA: Blackwell, 2005. Print.

Castellanos, A. et al. "Tapirus bairdii." 2008. *IUCN Red List of Threatened Species 2012.* Web. 28 March 2013.

Dawn of the Maya. Dir. Graham Townsley. National Geographic, 2005. DVD.

Faber, Daniel. *Environment Under Fire: Imperialism and the Ecological Crisis in Central America.* New York: Monthly Review Press, 1993. Print.

French, Jennifer. *Nature, Neo-Colonialism, and the Spanish American Regional Writers.* Hanover, NH: Dartmouth College Press, 2005. Print.

Heffes, Gisela. *Políticas de la destrucción / poéticas de la preservación: Apuntes para una lectura eco-crítica del medio ambiente en América Latina.* Rosario: Beatriz Viterbo Editora, 2013. Print.

Heyman, Gene M. *Addiction: A Disorder of Choice.* Cambridge: Harvard UP, 2009. Print.

Honey, Martha. *Ecotourism and Sustainable Development: Who Owns Paradise?* Washington, DC: Island Press, 1999. Print.

Kane, Adrian Taylor. *The Natural World in Latin American Literatures: Ecocritical Essays on Twentieth Century Writings.* Jefferson, NC: McFarland, 2010. Print.

Kearns, Sofía. "Nueva conciencia ecológica en algunos textos femeninos Contemporáneos." *Latin American Literary Review* 34.67 (2006): 111–27. Print.

———. "Otra cara de Costa Rica a través de un testimonio ecofeminista." *Hispanic Journal* 19.2 (1998): 313–39. Print.

Naess, Arne. "The Shallow and the Deep, Long-Range Ecology Movement: A Summary." *The Deep Ecology Movement: An Introductory Anthology.* Ed. Alan Drengson and Yuichi Inoue. Berkeley: North Atlantic, 1995. Print.

Norris, Ruth, J. Scott Wilber, and Luís Oswaldo Morales Marín. "Community-Based Ecotourism in the Maya Forest: Problems and Potentials," in Primack et al. 327–42. Print.

Paravisini-Gebert, Elizabeth. "Caribbean Utopias and Dystopias: The Emergence of the Environmental Writer and Artist." *The Natural World in Latin American Literatures: Ecocritical Essays on Twentieth Century Writings.* Ed. Adrian T. Kane. Jefferson, NC: McFarland, 2010. 113–35. Print.

Payan, E. et al. "Leopardus wiedii." 2008. *IUCN Red List of Threatened Species 2012.* Web. 28 March 2013.

Paz, Octavio. *Poesía en movimiento: México, 1915–1966.* Mexico City: Siglo XXI, 1966. Print.

Postema, Joel. "Ecology and Ethnicity in Anacristina Rossi's La loca de Gandoca." *Cincinnnati Romance Review* 27 (2008): 113–24. Print.

Primack, Richard B., David Bray, Hugo A. Galleti, and Ismael Ponciano, eds. *Timber, Tourists, and Temples.* Washington, DC: Island Press, 1998. Print.

Quiroga, Horacio. "El hombre muerto." 1920. *Cuentos.* Madrid: Cátedra, 2005. 308-12. Print.

Rey Rosa, Rodrigo. *Lo que soñó Sebastián.* Barcelona: Seix Barral, 1994. Print.

Rivera Barnes, Beatriz, and Jerry Hoeg. *Reading and Writing the Latin American Landscape.* New York: Palgrave Macmillan, 2009. Print.

Rossi, Anacristina. Author's Note. "Mad About Gandoca." Trans. Regina A. Root. *Interdisciplinary Studies in Literature and Environment* 15.1 (2008): 195–215. Print.

———. Interview. "True Fiction: Costa Rican Novelist Tells Stories that Need to be Told." By Carol Polsgrove. *Carol Polsgrove on Writers' Lives.* N.p. March 2011. Web. 9 Sept. 2011.

———. *La loca de Gandoca.* 1992. 9th edition. Editorial Legado: San José, Costa Rica, 2008. Print.

———. "RE: Una pregunta sobre La loca de Gandoca." Message to Adrian T. Kane. Trans. Kane. 12 June 2013. E-mail.

———. Skype Interview. 30 Sept. 2011.

Silvius, Kristen. "Forests." Latin America and the Caribbean: A Continental Overview of Environmental Issues." *The World's Environments.* Santa Barbara: ABC-CLIO, 2004. Print.

Thorbjarnarson, John B., Harry Messel, F. Wayne King, James Perran Ross. *Crocodiles: An Action Plan for Their Conservation.* Gland, Switzerland: International Union for Conservation of Nature and Natural Resources, 1992. Print.

Tittler, Jonathan. "Ecological Criticism and Spanish American Fiction: An Overview." *The Natural World in Latin American Literatures: Ecocritical Essays on Twentieth Century Writings.* Ed. Adrian T. Kane. Jefferson, NC: McFarland, 2010. 11–36. Print.

West, Robert. *Theory of Addiction.* Oxford, UK: Blackwell, 2006. Print.

White, Steven F. *Arando el aire: La ecología en la poesía y la música de Nicaragua.* Managua: 400 Elefantes, 2011. Print.

Williams, Raymond L. "Modernist Continuities: The Desire to be Modern in Twentieth-Century Spanish-American Fiction." *Bulletin of Spanish Studies* 79 (2002): 369–93. Print.

Zoomers, Annelies. "Globalisation and the Foreignisation of Space: Seven Processes Driving the Current Global Land Grab." *The Journal of Peasant Studies* 37.2 (2010): 429–47. Print.

Chapter Nine

Northern Ireland ↔ Global South

James McElroy

Where better to initiate a discussion of how to provincialize discourse (to use Lawrence Buell's phrase) than from right inside the United Kingdom where one of Britain's oldest and still extant colonial properties, Northern Ireland, is located? While it has become something of an a priori among some postcolonial theorists that Ireland was never really a colony, ergo, Northern Ireland is not a colonial entity at present, such theorists ignore the otherwise inconvenient truth that their new normal, as theoretical practice, substantiates rather than subverts that anomalous congerie known as "The United Kingdom of Great Britain and Northern Ireland."

Luke Gibbons, in a review of Ashcroft, Griffiths, and Tiffin's *The Empire Writes Back* (1998), directly questions such prepossessive postcolonial theories, making the point that Ireland and Irish culture "is conspicuous by its absence (or exclusion) from most theorizations of the postcolonial" (27). As a result of such absence, continues Gibbons, Ireland is often "subsumed into an undifferentiated Britishness as part of the expansion of empire" while such theoretical practices, themselves, participate in the colonization of academic research concerning Irish literature and culture (27). Gibbons then reminds his readers that even though Ireland might have become Anglophone—was, in undifferentiated racial terms, white—it nevertheless found itself "in the paradoxical position of being a colony within Europe" (27). Terry Eagleton contributes to this debate by asking point blank, and with immediate reference to Northern Ireland, "Why is it plausible not to see Northern Ireland as a colony?" (327). Eagleton's extended response involves getting down into the weeds of Irish politics and making the point that colonialism and its attendant theories often serve economic and political needs that are internecine, if not comprador, in character:

Like the Almighty, capital is now so universal as to be completely invisible, which is why many of the Irish are now devout political atheists, having ceased to believe in colonialism at all. It is important for some of the Irish in the Republic not to believe that the Republic is post-colonial, since this would imply that it was previously colonised, which might in turn suggest that Northern Ireland was still a colony, which might in turn have unpalatable political consequences. (328–29)

But what, one might ask, does all this have to do with ecocriticism, or, as Lawrence Buell now prefers, "environmental criticism" (88)? Well, quite a bit. Because if, as is argued here, Northern Ireland constitutes a colonial matrix then the North might be viewed as Ireland's real Global South—as compared to Ireland's postcolonial South—when it comes to understanding ecological conditions inside a colonially devised political state circa 2014/ 2015. What it also means is that if we ever want to begin reading Northern Ireland's literature from an ecocritical standpoint then some degree of proprietorial awareness will need to be brought to bear on the invariable, even if sometimes invisible, connections between colonial experience and environmental history. The same awareness might also serve as a reminder that most critics, and this includes Gibbons, Eagleton, et al., fail to recognize that the North's political circumstances are related to a unique set of environmental conditions where colonial suffering and ecological degradation have gone hand in hand.

Barry Sloan, by no means an ecocritic, draws an interesting connection between the politics of civil strife in Northern Ireland and the things of nature that prevail in the North when he writes that "Northern-based critics have done little to establish a 'totality' which relates the pentameter and outdoor relief riots, pastoralism and the B-Special constabulary" (64). The Northern Ireland of which Sloan speaks is, for the record, one of the least wooded regions in Europe with only about six percent tree cover. It is also, or at least was, the most industrialized part of Ireland, reflecting intimate colonial ties to Britain with Belfast playing host to the shipyards of Harland & Wolff, aircraft manufacturers like Short Brothers, and a host of other large industrial concerns including DuPont, ICI, British Enkalon.

The same North—as both industrial and political state, or statelet—has also featured discrimination, repression, and blatant gerrymandering as normative with the region embodying a stringent system of differences in the material conditions, both ecological as well as economic, between its English-Scottish (Protestant) colonists and its Irish (Catholic) natives.

For Donna L. Potts such differences, compounded by differences between those communities which most readily identified with industrial practices and those which were grounded in longstanding rural traditions, have come to emblematize the tensions between Catholics and Protestants with Catholics asserting, often in loco-descriptive verse, their inherent right to think about Northern Ireland in "precolonial" terms (79).

These irreducible eco-historical differences coupled with the sectarianization of economic, cultural, and ecological space are in the end responsible for determining two broad approaches to viewing nature and the environment as between a Protestant colonial class and a native Catholic population.

As Potts sees it, the fact that the North was industrialized to a much greater extent than the South means such industrialization is "felt more strongly in its pastoral tradition" (12). Potts adds that the differences between, as well as within, such "pastoral" modes of reference indicate that in the case of Northern Ireland "the strategic retreat of pastoral has often been the means for contextualizing the sectarian violence of the Troubles," while strategic retreat, in the South, has often been (she references the short-lived euphoria surrounding the Celtic Tiger) due to modernist forces "unleashed by the Irish themselves" (12).

For Marianne Elliott, any attempt to contextualize the North and its dividedness must first begin by acknowledging that sectarian consciousness was and remains a pervasive force in Northern Ireland (182). And while the aggregate expressions of such sectarian difference might sometimes be irregular or indistinct, such differences nevertheless persist in concentrated form and reflect/refract marked differences in colonial experience including the coextensive realities of eco-apartheid on the ground.

Tom Paulin, a Northern Protestant, addresses such sectarianism head-on in his poem, "Of Difference Does It Make" (1983), where he cites a rare example of bipartisanship in the North having to do with a single piece of environmental legislation. Paulin's poem, a broadside of sorts, is based on the fact that the only Bill to ever successfully emerge from a non-Unionist sponsor at Stormont, the North's predominantly Protestant parliament, was the Wild Birds Act of 1931 because bird conservation was deemed to be so inconsequential that sectarian difference did not matter one iota when it came to talking about plovers and stonechats or a rare (as in fictional) stint "called the notawhit" which is said to rap out "a sharp code-sign / like a mild and patient prisoner / pecking through granite with a teaspoon" (64).

John Hewitt addresses much the same sectarian issues—"sharp code-sign"—in his article, "The Bitter Gourd" (1945), where he recognizes that "Ulster's position in this island involves us in problems and cleavages for which we can find no counterpart elsewhere in the British archipelago" (109). And for Hewitt, at least, such "cleavages" are to be traced back, as he puts it in "The Course of Writing in Ulster," to the fact that Northern Ireland began as "a colony of mixed races in a hostile country" (76). While Hewitt does his level best to make a case for the North's literature as a regional literature, even he realizes, as he acknowledges in another critical piece, "No Rootless Colonist," that because "Planter stock often suffer from some crisis of identity, of not knowing where they belong" Protestant writing contains, within its own stringent terms of reference, an element of crisis that can never be eradicated given the colonist's standing within the North's social and political order (146).

Sam Robertson's "John Hewitt's Allegorical Imagination" argues that even some of Hewitt's best known nature poems—"The Stoat," "Hedgehog," "The Watchers," "The Blossomed Thorn"—contain frequent and persistent reminders that lodged deep inside the Northern Protestant poet's "gaze on nature lurk the tensions of identity politics in Ulster" (168). According to Seamus Heaney, these tensions underscore the fact that Hewitt is "in possession of another vocabulary and another mode of understanding" as compared to his own indigenous (as it were) sense of nature and the natural world (147). And it is this difference, says Heaney that registers a distinctive ecological perspective which has its source in Hewitt's unique colonial inheritance: "his cherishing of the habitat is symptomatic of his history, and that history is the history of the colonist" (147).

Hewitt himself recognizes, in "Once Alien Here"(1943), that his settler ancestors were responsible for taming, as he likes to put it, the wilderness that defined ancient Ireland: "Once alien here my fathers built their house, / claimed, drained, and gave the land the shapes of use" (20). He also recognizes that his own people favor picturesque versions of the North's countryside, as is the case in "The Glens," where he writes, "My dead / lie in the steepled hillock of Kilmore / in a fat country rich with bloom and fruit" (54).

As much as Hewitt might try his level best to fight a rearguard action when it comes to defining, and defending, the sectarian eco-differences that have held sway in the North since the 1930s, the truth is that most Northern Protestant poets, both then and since, have tended to deal with their "crisis of identity" by distancing themselves from the region's eco-exigencies or addressing environmental issues from somewhere outside Northern Ireland.

Louis MacNeice, for one, spends little or no time trying to make a case on behalf of the natural order found in Northern Ireland but prefers to describe the North, if and when he describes it at all, as an ecological backwater with lingering criticisms of Belfast featured in the poem "Belfast" (1931): "Down there at the end of the melancholy lough / Against the lurid sky over the stained water / Where hammers clang murderously" (17). MacNeice's critique of Belfast is extended in "Carrickfergus" (1937) where, after he tracks residential sectarian difference—"The Scotch Quarter was a line of residential houses / But the Irish quarter was a slum"—he again criticizes the city and its environs for serving as little more than an elaborate toxic waste site where it is common to see a brook run "yellow from the factory stinking of chlorine" (69).

Beyond his criticisms of Belfast, or rather because of his criticisms of Belfast, MacNeice often takes refuge in the West of Ireland while the North and a lot of its variegated (other) habitats are kept under wraps. This retreat into an Ireland other than the North is, according to Edna Longley, a characteristic stratagem of Northern Protestant writers who embrace the picturesque wildness of "the West" that much more than their Catholic counterparts (110). It should come as no great surprise, then, that MacNeice situates his natural world amidst "the wilds" of the West. In contrast to the untamed landscapes and seascapes of the West, MacNeice makes clear that he, like his father, finds "the English landscape tame" and proceeds to enter a poetic critique of the enclosure England represents, saying that England's "woods are not the Forest; each is moored / To a village somewhere near" (231). Whatever about the fact that the wilds of Ireland's West represent, for Mac-Neice, a place apart—as it did for Anglo-Irish Protestant writers like W. B. Yeats and John Millington Synge before him—it is important to remember that MacNeice's Mayo, as host environment, is absent the interposing presence of any real human inhabitants, human habitation, or human involvement with reference to the landscapes he poeticizes.

Following in MacNeice's footsteps, Michael Longley places a large portion of his work in and around Carrigskeewaun in County Mayo. Tom Herron argues, and with good reason, that the ecological relationships Longley establishes in his Carrigskeewaun poems neglect "local intra-human contact, as if human beings were somehow exempt from ecological consideration" (80–81). Herron adds that one of the most interesting, and revealing, things about Longley's poetry is that it has a "remarkably deterritorialized relation to land and landscape" (83). As often as Longley might therefore be featured as a celebrant of nature there remains a certain absence at the heart of his nature narratives whereby Northern Ireland is often left out of account while reference to his adopted homeland in "the wilds" of Mayo is presented as a safe haven (non-colonial realm) in such poems as "Carrigskeewaun" (68–69), "Skara Brae" (71), and "On Mweelrea" (142).

Longley once said, in an interview with Jody Allen Randolph—and this in the face of civil strife in the North—that "The most urgent political problems are ecological: how we share the planet with the plants and other animals. My nature writing is my most political" (298). It is interesting to note that in the same interview Longley insists he is not "trying to escape from political violence" by writing about Mayo but is, instead, searching for ways to get "the light from Carrigskeewaun to irradiate the Northern darkness" (305). Ironic or not, this last reference to his poems irradiating the North's darkness serves as yet another curved reminder, to quote Herron, that Longley's West "provides an alternative ecology to that subsisting in the North" (80).

The point here, of course, is not so much that, but why, Longley and other Protestant poets find it so difficult if not impossible to narrativize the environment of the North they grew up in rather than irradiating it from afar. Put another way, why does so much of Longley's eco-rhetoric neutralize or negate the natural order that holds sway in the North? Derek Mahon, who provides as much, if not more, poetic comment than Longley on a wide range of ecological issues such as climate change, habitat loss, pollution, deforestation, etc., revisits, in "Autumn Skies" (2010), the question of just how much, as Protestants from the North, he and Longley both experience a certain "cultural confusion" (akin to Hewitt's "crisis of identity") that cries out for some kind of resolution—the kind of resolution that Mahon thinks Longley may have found in a remote "cottage down the west / and took the answer back / to battered old Belfast" (355–56).

Mahon's own brand of eco-politics, especially in recent collections, is often situated at nominal sites outside the North. And while it is true that he sometimes provides available references to Northern Ireland's standing topographies interspersed with references to plants, birds, and so on, he nevertheless tends to keep his distance from the North as an eco-political province and all the problems that lie therein. In fact, on those occasions when Mahon does spend some time writing about the destruction of native Irish habitat his environmental concerns are, more often than not, located in an Ireland other than the North: where MacNeice and Longley sought, and to some extent found, a measure of cultural and ecological solace in the so-called wilds of Mayo, Mahon finds something similar in what he calls, "After the Storm" (2010), his own secluded corner of "quiet Munster" (344–45).

Rajeev Patke, in "Partition and its Aftermath: Poetry and History in Northern Ireland" (2010), writes that Northern Ireland's poets try "to marginalize the politics of partition by practicing a poetics of obliquity" (24). This is especially true of poets who hail from the North's Protestant tradition as evidenced by the aforementioned narratives where Northern Ireland, as ecological subject, is consigned to a condition of virtual unpresence: a location approximating to a kind of elsewhere that stands in marked contrast to the narratives of native (Catholic) poets like John Montague who in his evocation of Garvaghey—Garvaghey comes from the Irish, garbh acaidh, which translates as "the rough field" in English—celebrates the subsistence terrains that disenfranchised Catholics were often forced to inhabit in the North.

Elmer Kennedy-Andrews makes the point that Montague has always been careful to reject both narrow regionalism and "boundless" globalism in his poetic representations of Garvaghey and its surrounds (31). Kennedy-Andrews also makes the point that Montague rejects "the bogus pastoral of the Irish past" while his search for some kind of center, or source, "is conducted in full awareness that the idea of a centre is fictive" (38). According to Potts, this rejection of bogus pastorals serves as a valuable reminder that Montague is keenly aware of the fact that his native habitat is in "shards as a result of colonization" (30). Acknowledging, as he does, that his own townland is different from those landscapes identified with certain English traditions—

"No Wordsworthian dream enchants me here"—Montague draws an imposing contrast in the first section of his long poetic tract, The Rough Field
("Home Again"), between non-native and/or Protestant vistas and the wine
country to which he and his fellow Catholics belong: "merging low hills and
gravel streams, / Oozy blackness of bog-banks, tough upland grass; / Rough
Field in the Gaelic and rightly named" (9).

This "Harsh landscape that haunts me," as Montague puts it later in the
same poem, is therefore something—to use a modish term—that he "owns."
Set, as it is, amidst "bog-banks," "tough upland grass," "merging low hills,"
and "gravel streams," Montague uses his townland's eco-identifiers to explore the fate of native species in poems like "Springs" (1989) where he
recounts the suffering of a salmon which Potts is convinced (44) bears all the
diagnostic traces of UDN, or ulcerative dermal necrosis, with sores surrounding the fish's gills and eyes (296). By doing this, Montague ensures that the
poem's ailing salmon bears witness not only to its own condition but to a host
of other changes that have taken place in his ancestral homeland wherein he
records "The disappearance of all signs / of wild life" (56) and serves up a
grim poetic précis about what progress, measured in purely human terms,
signals for the future of Garvaghey as a distinct bioregional location.

Such disappearance of wild life also involves the destruction of nearby
habitat formations when "A brown stain seeps away from [. . .] the machine"
with local grasses becoming discolored and the nearby trout stream "thicken-
ing" with effluent. Taken together, these and several other pointed references
to "brown stain" and "broken banks" are nothing if not sustenant eco-mark-
ers that help distinguish between the harsh landscapes of Montague's Garvaghey and the solicitous pastorals of John Hewitt's Antrim Glens or the trans-
posed eco-scenarios of the West/South-West of MacNeice, Longley, and
Mahon. All of which makes perfect sense if we stop to consider that there is a
virtual absence of any raw, harsh, or bog-grained references in most Northern
Protestant texts. It also makes perfect sense if we stop to consider Catherine
Wynne's position, set out in "The Bog as Colonial Topography in Nineteenth-Century Irish Writing," which "The figure of the bog operates as an
agent of disruption, with shifting and ambiguous social and political alle-
giances that simultaneously mirror and mask the instability of the colonised
topos" (322).

One of the most famous exponents of the "bog" poem is Seamus Heaney
who has used bog references for political, archaeological, ecological, and
other reasons in order to facilitate occasions of potential hermeneutic contact
between man and nature. For Heaney, as for Montague, bogs and their kin-
dred habitats represent irreducible signifiers having to do with limited means
and neglected holdings where, as Heaney lines it in "Serenades" (1972),
someone might come upon the "ack-ack" of a corncrake "Lost in a no-man's-
land / Between combines and chemicals" (72). Heaney uses the same repre-

sentational eco-borders in the poem, "Bogland," to help differentiate between his own family's holdings and the greater breadth of alternative environments (on this particular occasion the unfenced prairies of the United States) because for those who live and work in the North that Heaney identifies with any discussion of habitat—intransigent native habitat—involves an acute awareness that "Everywhere the eye concedes to / Encroaching horizon" (41).

Alan Gillis, who hails from a Protestant background and belongs to a more recent generation of Northern poets, wants nothing to do with Heaney's meditations on the rural with its "Encroaching horizon." Gillis makes this altogether clear in "After Arcadia," from *Hawks and Doves* (2007), where he writes, "Long since the cuckoo and curlew, nightingale and night- / jar have been silenced; the suburbs have long seized every valley" (57). But it is in "The Ulster Way," from *Somebody, Somewhere* (2004), that Gillis issues his most categorical statement about banishing Ireland's rustic imperative because his poetic is most certainly "not about horizons, or their curving / limitations" (9). According to Gillis, then, Northern Ireland's poets should renounce all things rural including the botanies of "gorse" (Heaney prefers the more native reference to "whin") and tells his readers, straight out, that they "will not be passing into farmland" since any consideration of rustic habitational space sets unacceptable limits on what modern Northern poets could, or should, write (9).

Matt McGuire, reinforcing the basic premise that Gillis's work is urban in orientation—"infused with the inevitable entropy of post-millennial culture" (89)—is quick to add that his "antinomy to the rural landscape signals a deeper dissatisfaction" because representations of the rural landscape In Northern Ireland have often been used to "imbue the ideologies of nationalism and unionism" (91).

While Gillis clearly has every right to question the limitations of the rural poem and the ideological risks involved in using such poetic forms, it is worth noting that in rejecting Heaney's habitational horizons (wholesale) he belies any and all prospective discussions of alternate eco-poetic practices as well as circumventing a few awkward questions about the status of Northern Ireland as a territorial eco-holding where colonial usurpation and environmental degradation have left their mark.

Sinéad Morrissey, who hails from a Catholic background and is a contemporary of Gillis's also provides alternative narrative strategies to those of Heaney though she, unlike Gillis, is prepared to engage with the North's environment as a subject of some importance. This is something Morrissey makes altogether clear in *Through the Square Window* (2009) where she, like some of her forebears, offers—though she is far from being a rustic poet— a sustained poetic measure of the environmental issues facing Northern Ireland today. One example of such engagement is her "Cycling at Sea Level" where she talks about urban development in and around Belfast's Duncrue Industrial Estate "whose meat-/plant and meal-factory have threaded the air with dust" (22). Morrissey also talks about Belfast Dump's Shore Park where she catalogues an industrial landscape "looking for all the world / as though some meteor hit and killed off half the planet" (22). With "The Hanging Hare" (2009) Morrissey continues her assessment of the North's natural world juxtaposing, as she does, the human debris that once surrounded her childhood home—"the sunken playground, hacked-out stumps of trees / & blackened mattresses where a fire had been"—with the outside possibility that such squalid conditions could be restored to something much more eco-logically sound if people embraced Ireland's legacy of "unimpeded air" and its surviving memories habitat for a wild hare (52).

Morrissey thus feels entitled to author an encomium for the North's near-extinct "spaces" (52). In so doing, Morrissey engages in modest acts of eco-reclamation while her Protestant contemporaries, Gillis among them, contin-ue to experience an unbidden though abiding "crisis of identity" when it comes to writing about environmental issues and concerns inside the geo-political borders that make up Northern Ireland. There is in truth something eco-dissociative at work in all this and unless modern theorists are prepared to recognize that Northern Ireland is a de jure colony then their failure to do so (coefficient theories notwithstanding) will continue to sanction a brand of critical analysis which never comes to grips with just how much colonialism as process, province, and product is able to make itself, and its eco-systems, illegible in the midst of the legible.

Perhaps one of the most intriguing facets of this "illegibility" is how extant post/colonial conditions influence the ways in which theoreticians envisage the colonial as colonial—or not. The same influence, an inadequate term under the circumstances, also has some bearing on the degree to which a number of theorists have been blinded to the otherwise obvious proposition that Northern Ireland is founded on hardcore principles of eco-colonial segre-gation wherein its colonial settler classes (Protestant) feel compelled to write about the North's natural environment from inside a seemingly acolonial eco-space while its native-indigenous classes (Catholic) feel compelled to assert a kind of irredentist right to own it all even when "it" is considered to be an integral part of "The United Kingdom of Great Britain and Northern Ireland" and so is presumed to have no links, no links whatsoever, to the eco-diasporas that inhabit what is now termed the Global South.

WORKS CITED

Buell, Lawrence. "Ecocriticism: Some Emerging Trends." *Qui Parle: Critical Humanities and Social Sciences* 19.2 (2011): 87–115. Print.

Eagleton, Terry. "Afterword: Ireland and Colonialism." *Was Ireland a Colony? Economics, Politics and Culture in Nineteenth-Century Ireland*. Ed. Terrence McDonough. Dublin: Irish Academic Press, 2005. 326–33. Print.

Elliott, Marianne. "Religion and Identity in Northern Ireland." *The Long Road to Peace: Peace Lectures from the Institute of Irish Studies at Liverpool University*. Ed. Marianne Elliott. 2nd ed. Liverpool, UK: Liverpool UP, 2007. 175–91. Print.

Gibbons, Luke. "Ireland and the Colonization of Theory." *Interventions: International Journal of Postcolonial Studies* 1.1 (1998): 27. Print.

Gillis, Alan. *Hawks and Doves*. Loughcrew, Oldcastle, UK: Gallery Press, 2007. Print.

———. *Somebody, Somewhere*. Loughcrew, Oldcastle, UK: Gallery Press, 2004. Print.

Heaney, Seamus. *Preoccupations: Selected Prose 1968–1978*. London: Faber and Faber, 1980. Print.

———. *Opened Grounds: Poems 1966–1996*. London: Faber and Faber, 1998. Print.

Hewitt, John. *The Selected John Hewitt*. Ed. Alan Warner. Belfast, Ireland: Blackstaff Press, 1981. Print.

———. *Ancestral Voices: The Selected Prose of John Hewitt*. Ed. Tom Clyde. Belfast, Ireland: Blackstaff Press, 1987. Print.

Kennedy-Andrews, Elmer. "John Montague: Global Regionalist?" *The Cambridge Quarterly* 35.1 (2006): 31–48. Print.

Longley, Edna. *Poetry & Posterity*. Tarset, Northumberland, UK: Bloodaxe Books, 2000. Print.

Longley, Michael. *Collected Poems*. Winston-Salem, NC: Wake Forest UP, 2007. Print.

———. "Interview: Michael Longley and Jody Allen Randolph." *Colby Quarterly* 39.3 (2003): 294-308. Print.

MacNeice, Louis. *The Collected Poems of Louis MacNeice*. Ed. E.R. Dodds. New York: Oxford UP, 1967. Print.

Mahon, Derek. *New Collected Poems*. Loughcrew, Oldcastle: Gallery Press, 2011. Print.

McGuire, Matt. "Northern Irish Poetry in the Twenty-First Century." *No Country for Old Men: Fresh Perspectives on Irish Literature*. Ed. Paddy Lyons and Alison O'Malley-Younger. Bern, Switzerland: Peter Lang, 2009. 87–101. Print.

Montague, John. *Collected Poems*. Winston-Salem, NC: Wake Forest UP, 1995. Print.

Morrissey, Sinéad. *Through the Square Window*. Manchester, UK: Carcanet, 2009. Print.

Patke, Rajeev. "Partition and its Aftermath: Poetry and History in Northern Ireland." *Journal of Postcolonial Writing* 46.1 (2010): 17–30. Print.

Paulin, Tom. *Selected Poems 1972-1990*. London: Faber & Faber, 1993. Print.

Potts, Donna L. *Contemporary Irish Poetry and the Pastoral Tradition*. Columbia: U of Missouri P, 2011. Print.

Robertson, Sam. "John Hewitt's Allegorical Imagination." *Irish Studies Review* 17.2 (2009): 167–82. Print.

Sloan, Barry. *Writers and Protestantism in the North of Ireland: Heirs to Adamnation?* Dublin, Ireland: Irish Academic Press, 2000. Print.

Wynne, Catherine. "The Bog as Colonial Topography in Nineteenth-Century Irish Fiction." *Was Ireland a Colony?* Ed. Terrence McDonough. Dublin, IRELAND: Irish Academic Press, 2005. 309–25. Print.

Chapter Ten

"Decline and Fall": Empire, Land, and the Twentieth-Century Irish "Big House" Novel

Eóin Flannery

How should the world be luckier if this house,
Where passion and precision have been one
Time out of mind, became too ruinous
To breed the lidless eye that loves the sun?
—William Butler Yeats, "Upon a House shaken by the Land Agitation" (77)

I

Ireland's status as a postcolonial society has consistently been contested within the fields of Irish historical and Irish cultural studies, as well as often being overlooked within the discipline of international postcolonial studies. In addition, Irish political and cultural histories have only recently been approached in substantive ways using the resources of ecocriticism. This is despite the fact that landscape was a central and enduring part of the imaginaries of British colonialism in Ireland and of anti-colonial Irish nationalisms. In a recent essay on Ireland and the British Empire in the nineteenth and early twentieth centuries, Alvin Jackson crystallizes one of the major contradictions that informed the Irish–British colonial relationship. He argues that:

> Ireland was simultaneously a bulwark of the Empire, and a mine within its walls. Irish people were simultaneously major participants in Empire, and a significant source of subversion. For the Irish the Empire was both an agent of

liberation and of oppression: it provided both the path to social advancement
and the shackles of incarceration. (123)

Ireland's geographical proximity to mainland Britain nourished a sense of
unpredictable intimacy between the two islands, a state of affairs that was
directly contributory to the historical designs to colonize and subdue the
island. Very real security risks were posed to the British mainland by the
potential use of Ireland as a base for French military aggression. It is therefore understandable that Irish colonial history is marked by repeated military,
political, constitutional, and confessional endeavors to bring the 'unruly'
island to colonial subjection.

A presiding concern of the British colonial polity was the need conclusively to assimilate all of its Celtic peripheries, including Ireland, Scotland,
and Wales. The sustained colonization of Ireland began in the sixteenth
century and continued into the early seventeenth century with a series of
settler plantations in the south, east and north of the island. It was a period of
conquest and settlement that, as Jane Ohlmeyer concludes, involved: "strategies [which] though often couched in the rhetoric of civility, effectively
amounted to a form of imperialism that sought to exploit Ireland for England's political and economic advantage and to Anglicise the native population" (28–29). However, when we reach the eighteenth century, the constitutional countenance of Ireland has altered. By this period Ireland had become
a formal kingdom and was possessive of its own parliament—a fact that not
only differentiated it from contemporary, and many subsequent, British colonies, but supplements the catalogue of contradictions that besets Irish colonial history. We shall see that the cosmetics of constitutional parity of esteem
too often mask the endurance of colonial subjugation.

Further dispute arises when we confront Irish history in the nineteenth
century, in particular after the passing of the Act of Union in 1801. Under the
Act, which was legislatively operative until 1921, Ireland was ostensibly a
partner in the United Kingdom. Thus, any argument that suggests Ireland
was a bona fide colony of the British Empire is apparently confounded by
recourse to Ireland's de facto legal-constitutional status. However, the actual
equity of this constitutional partnership is, at best, questionable. While equal-
ity may have been legally enshrined through the Act of Union, Ireland retained many of the political characteristics of a colonial administration. Primary among these accretions of colonial authority were the retention of both
a Chief Secretary and a Lord Lieutenant in Dublin, together with the maintenance of an executive in Dublin Castle. In addition to these procedural indi-
ces of colonial administration, Jackson alludes to more affective and revealing features of the British dispensation in Ireland. He notes: "English officials billeted in Ireland developed the same attitudes of mixed bemusement,

condescension, affection, and eagerness to help which characterized their counterparts in India and elsewhere" (126).

Jackson opens up a crucial field of debate within postcolonial studies. He suggests the extent to which a reductive reliance upon empirical facts too often conceals the affective peculiarities of intercommunal and interpersonal relations within colonial societies. Despite the apparent constitutional parity granted to Ireland as part of its union with Britain, many contemporary Irish cultural critics and historians readily affirm its historical condition as that of a colonized society. This diagnosis is founded on readings of the impact of imperial modernity on Ireland. It underwent intense and uneven experiences of modernization via colonialism, rather than under the processes of a modulated history of industrialization. The most consistent preoccupation of Irish postcolonial studies is the historical exposure of the discursive mechanisms through which Ireland was represented as imperial Britain's, and by extension modernity's, recalcitrant "other within." Regardless of the patina of legal union that obtained in the nineteenth century, Ireland remained culturally, confessionally, and economically recalcitrant to the civilizational calculus of imperial modernity.

II

Looking at one subgenre of Irish literary history—the twentieth-century "Big House" novel—this chapter will critically read a selection of texts that draw upon its templates. The novels to be considered in this discussion are: *The Last September* by Elizabeth Bowen (1929); *Troubles* by J. G. Farrell (1970); and *The Newton Letter* by John Banville (1982). Combining close textual analysis with relevant historicization, the chapter will illuminate the ways in which these differentiated legatees of the nineteenth-century "Big House" novel represent the hubris and the decline of a semi-feudalist colonial system in Ireland. In dealing with this subgenre we will, firstly, detail, in brief, the history and the features of the "big house" cultural economy in Ireland since the late eighteenth-century, and discuss how this system was integral to the British colonial occupation of Ireland. I will argue that they belong to either the end of the first generation of the subgenre, or to a second generation that utilises the subgenre to reflect upon the contested legacies of Irish [colonial] history and the writing of history. The focus on Irish history, which is common to all generations of the Irish "Big House" novel, will also permit a discussion of the subgenre's relationship with the Gothic.

John Banville's 1982 novel, *The Newton Letter*, retrospectively details the professional, philosophical and personal struggles of a historian of Isaac Newton's life and career. These struggles are not mutually exclusive in the narrative, as Banville's narrator navigates between lucidity and misrecognition in his historical travails, as well as in his sexual intimacies in the present. The historian is a temporary tenant of a lodge on a landed estate, though the narrative is set in the contemporary moment. This setting allows Banville to confront the politics of historical representation, and of epistemology generally, as his protagonist fails to grasp the 'true' nature of the reality he is ensconced in at the estate at Ferns. For our purposes, Banville's self-reflexive and, to many, postmodern narrative, reprises some the generic codes of the historical "Big House" novel. Philosophically, the gap between the narrator's assumptions about the resident family's status, and the facts of their existence on this ailing estate, throw into relief the disparities between representation and reality. Opening the narrative with an address to an absent correspondent, the narrator confesses: "I have abandoned my book [. . .] I've lost my faith in the primacy of text [. . .] Real people keep getting in the way now, objects, landscapes even" (1).

Though his primary motivation for his residency at Ferns was to complete a study of Newton, the history that impinges upon his consciousness most stubbornly is that of Ireland. The period during which these three imbricated crises peak is experienced within the confines of a primary colonial crucible, the landed estate. Indeed, prior to his arrival at Ferns, as he reveals, the narrator was laden with preconceptions about his destination: "I shouldn't have gone down there. It was the name that attracted me. Fern House! I expected—Oh, I expected all sorts of thing. It turned out to be a big gloomy pile with ivy and peeling walls and a smashed fanlight over the door, the kind of place where you picture a mad stepdaughter locked up in the attic" (3). As with the earlier confession, there is a tense dialogue between materiality and abstraction—the name of the "Big House," the signifier, is one of the causal factors at the outset of the narrator's journey. But equally noteworthy is the manner in which his pre-conceived fantasy of the house is explicitly extracted from a literary precursor, Charlotte Brontë's *Jane Eyre*, together with its metafictional legatee, Jean Rhys's *Wide Sargasso Sea*. Thus one strain within the narrative locates the family and the "Big House" within international colonial and literary historical contexts, yet the narrator's misconceptions are further revealed, at length, as 'received' from an Irish perspective:

> I had them spotted for patricians from the start. The big house, Edward's tweeds, Charlotte's fine-boned slender grace that the dowdiest of clothes could not mask, even Ottilie's awkwardness, all this seemed the unmistakeable stamp of their class. Protestants, of course, landed, the land gone now to gombeen men and compulsory purchase, the family fortune wasted by tax, death, duties, inflation. But how bravely, beautifully they bore their losses! (Banville 12)

In one respect, we see Banville playfully engaging with an established genre within Irish literary history, but at a textual level we witness his narrator structuring his own material experience in terms of "genre" also. The narrator's engagement with this family, this history, this ecology of conquest and postcolonial decline is fashioned from stereotype and conventional knowledge. He is seduced by the faded grandeur of his hosts and allows this fantastic seduction to "author" a version of family history remote from the material exertions of colonial settlement and expropriation. Despite the postmodern strategies of Banville, and the postmodern philosophizing of his protagonist, the present absences of colonial usurpation, of the colonial conjuncture detailed below by Eamonn Slater and Terrence McDonough cannot be entirely overwritten or erased. Banville's metafictional deployment of the "big house" in the late twentieth-century cannot avoid a confrontation with Ireland's colonial history. But in the current context, the text reminds us that literature, the novel, and, more precisely, the "big house" novel was just one of the modes through which attempts were made to hold a politically and culturally recalcitrant indigenous Irish population in representational "check." As well as entailing the violent seizure of human and nonhuman environments, Empire is discourse. The novels under consideration in our discussion dramatize these features of the Irish colonial experience, but they also reveal fractures in the physical and representational composure of imperial incumbents.

III

"I intend an emphasis," Raymond Williams writes in *Culture and Materialism*,

> when I say that the idea of nature contains, though often unnoticed, an extraordinary amount of human history [. . .]. A considerable part of what we call natural landscape [. . .] is the product of human design and human labour, and in admiring it as natural it matters very much whether we suppress the fact of labour or acknowledge it. (67, 78)

Williams's comments here are instructive because they underscore the extent to which the reified categories of "nature" and "culture" are too often aggregated in ecocritical and ecopolitical discussions. For Williams a critical materialist engagement with the politics of environmental destruction and exploitation cannot proceed without an appreciation of the "always-existing interaction between humans and environment, history and nature" (Mukherjee 61). Mukherjee's analysis is part of a broader skein of contemporary Marxian ecocritical interventions that draw upon Williams's proto-ecocritical writings. But Mukherjee is also attentive to, and keen to foster, the historical and contemporary mutualities of postcolonial and ecological criticisms. And it is at this critical junction that our analysis of the twentieth-century

Irish "big house" novel is sited; unpacking literary [cultural] narratives in which land appropriation; inheritance; class conflict; and economic disparity are dominant themes.

In their Marxist analysis of the feudalist colonial conjuncture that obtained in Ireland until the close of the nineteenth-century, Eamonn Slater and Terence McDonough begin by providing historical and statistical contexts for their arguments.[1] The Irish landscape, Irish society and the Irish economy were, in their view, conditioned by the legacies of the early modern colonization of the island. As they suggest: "Since the plantations of the sixteenth and seventeenth centuries, the land surface of Ireland has been divided into landed estates [. . .]. Some estates were as small as 100 acres but it was the large landed estates which predominated" ("Bulwark of Landlordism and Capitalism" 67). Perhaps the most startling statistic they offer is that by 1870 "only 3 percent of the population of the country owned land" ("Bulwark of Landlordism and Capitalism" 67). The historically protracted and geographically lateral colonization of Ireland, then, resulted in a deeply uneven and hierarchical social formation, particularly in rural areas, where the vast majority of the indigenous Irish population subsisted across the nineteenth century. In ecological terms, what we witness is a social metabolic rift as a consequence of which profits and incomes are heavily weighted in favor of the landlord class. While the labor of the tenanted classes increasingly service the needs of a proprietorial constituency. Not only was there a virtual monopoly on land ownership in nineteenth century Ireland, and earlier, but the Irish tenant class had few if any legal rights under this socio-economic settlement. And again, Slater and McDonough are illuminating on this disenfranchised legal status: "The bastardized English feudal system in Ireland failed to introduce any customary rights to the Irish feudal tenantry. In consequence, no tenurial customs were legally recognized. If they did exist, or develop, it was outside the legal system" ("Bulwark of Landlordism and Capitalism" 75). Once more the roots of legal inequality are historical, and they are consequential to the colonization of Ireland:

> Since the British conquest of Ireland, the landlord class was able to manipulate the Irish legal system in such a way that the tenantry was almost completely subjected to the arbitrary power of the landlord class. This was unlike the legal situation on the mainland in that the British tenantry were protected by Common Law. ("Bulwark of Landlordism and Capitalism" 75)

In a more recent intervention, "Marx on nineteenth-century colonial Ireland," Slater and McDonough detail Marx's thoughts on the 'Irish question' in his 1867 report on Ireland. Slater and McDonough re-iterate their earlier points on the legal inequities faced by the Irish tenant, but they invoke religious faith and environmental degradation in this later Marxist critique of

British colonialism in Ireland. Foregrounding the fractious relations that existed between landlord and tenantry, the authors note that "strategic necessity forged a string bond between Irish landed and British imperial interests." As a result of this, "The absence of traditional ties left the peasantry without customary protections. The religious disability of the peasantry reinforced inequity in tenurial relations [. . .]. Thus, the colonial character of Irish agricultural relations advantaged the landlords in conflict with both the native peasantry and alternative elites" ("Marx on nineteenth-century colonial Ireland" 172). Religious faith was a key structural feature of British colonialism in Ireland, and, as these comments highlight, Catholic peasant tenants were routinely exposed to the exploitative vagaries of a largely Protestant, and often absentee, landed gentry. The worst deprivations of the British feudal colonial system in Ireland were, then, endured by those without fixed rents or tenurial security. But this dominant socio-economic formation also had degrading effects on the nonhuman environment, a point noted by Marx, and underscored by Slater and McDonough: "Marx raised the possibility that the dynamic of colonial relations could push up against the limits of the physical environment. Colonial exploitation had the possibility of disrupting the reproductive processes involved in the recuperation of the soil through the recycling of essential nutrients" ("Marx on nineteenth-century Ireland" 172). In the Irish colonial context, what is significant here is Marx's co-implication of colonial occupation and the impacts on the physical, natural world. Indeed, from our current vantage point it is, perhaps, impossible to ignore the separation or mutual exclusivity of empire and ecological devastation. From Slater and McDonough's analyses, we glean a sense of the material social and economic realities of Ireland's feudal colonial formation—a formation that is the locus of the ensuing literary explications.

If the landed estate was the dominant unit of property under British colonialism in Ireland since the early modern period, then within each of these units the presiding seat of "local" authority "was the landlord's country residence, traditionally referred to in Ireland as the "Big House" by the wider community" (The Big Houses and Landed Estates of Ireland 9). The term "big house," then, can be read in both literal and pejorative terms, depending on the context. For Dooley, the nomination quite simply "captures the very essence of the raison d'etre behind the building of these houses—to announce the economic and social strength of their owners in their localities and as a class as a whole, to inspire awe in social equals and possibly encourage deference in the lower classes" (*The Big Houses and Landed Estates of Ireland*). The "big house" is, for all intents and purposes, an edificial correlative of the economic and political power attached to the landed gentry. Yet, as we shall see, and to contradict Dooley's final assertion above, such concreted grandeur could also be interpreted as a symptom of insecurity, transience, even an expression of a submerged guilt at the historical expropriations of this socio-economic class. When we begin to untangle the history and, for our purposes, the literary history of the "big house" and landlordism in Ireland, we unearth some provocative thematic complexes: power and vulnerability; belonging and alienation; loyalty and betrayal; and trust and paranoia. And Dooley portrays such an isolated and insulated class resident within the bounded expanses of the demesne and the "big house." He suggests: "Long before the 1870s, the typical Irish big house had become akin to an artificially created island. By reasons of wealth, social standing, religion, cultural upbringing, and political power, landlords and their families had become psychologically distanced from the vast majority of the people" (*The Decline of the Big House in Ireland* 18). While the "big house" stood as a physical and symbolic embodiment of a feudal colonial system, the surrounding demesne lands had their own discrete and interlinked ecologies. The landed estates that subdivided the colonized territory of Ireland imprinted not only on the "national" ecology, but also on localized ecologies, as Dooley describes: "all demesnes had much in common: demesne walls stretching for miles [. . .] gate lodges; plantations; open parkland; ornate gardens and expanses of lawn for recreational purposes; and kitchen gardens for the more practical purpose of keeping the big house self-sufficient" (*The Decline of the Big House in Ireland* 40).

IV

The protracted and traumatic nature of Ireland's colonial history, with its attendant litany of dispossession, atrocity, and sectarian division, furnished Gothic authors with a steady diet of atavistic antagonisms and malevolent mythologies. Even a cursory survey of the stylistic cosmetics of its literary conventions will confirm that inheritance, succession, and primogeniture fuel the suffocating paranoia of the tyrannous, the patriarchal and the aristocratic within the Gothic aesthetic. Regardless of the testimony of wills, legacies, and antique documents, despite such legislative undergirding, the question of legitimacy continually resurfaces. The contested nature of the Irish geographical and cultural landscape meant that these topographies were haunted by the disinherited revenants of colonial misappropriation. This colonial history was a fraction of what Mukherjee characterizes as "a permanent state of war on the global environment," conducted by "colonialisms and imperialisms" (68). The incendiary nature of history in Ireland created a landscape in which all history was personal. Any romantic sanitization of Ireland's rugged topography for the purposes of tourism belied the fractious memorial inheritances of the country's disenfranchised population. Segueing from the thematic logic of the Gothic genre to the historical binaries of reason and madness, modernity and tradition, and civility and barbarism, Siobhán Kilfeather suggests that "Gothic fiction is the appropriate self-image of enlightenment, because it plays with representations of domination, and represents the domination of our fears. It comes to terms with technological advancement and rational organization by opening a space for terror, regression and atavism" (36).

When we broach the Gothic in the Irish context, it is necessary to complicate our definitions of the term and of the genre. While English Gothic can justifiably be said to have a 'tradition', this is not the case in Irish literary history. Addressing the distinctions between its Irish and English variants, W. J. McCormack avers in his survey of Irish Gothic:

> Whereas the origins of English Gothicism are diverse and obscure—involving the sensibility of a remarkable individual like Walpole, the larger development of literary romanticism, and the growth of aesthetics through concepts such as "the sublime"—Irish Gothic fiction is remarkably explicit in the way it demonstrates its attachment to history and to politics. (833)

Not only are the origins of Irish Gothic trained on the [perceived] injustices of Ireland's colonial relationship with England, but within Irish literary history: "it has to be said that while Irish Gothic writing does not amount to a tradition, it is a distinctly protestant tradition" (McCormack 837). Thus McCormack's précis of Irish Gothic establishes a continuity with that of its English counterpart: its confessional allegiance. And this is a fact expanded upon by Terry Eagleton, who re-asserts the Protestant background of the Irish Gothic lineage, but develops upon McCormack's case by enlisting the peculiarities of Protestant experience in Ireland during the nineteenth-century. For Eagleton: "if Irish Gothic is a specifically Protestant phenomenon, it is because nothing lends itself more to the genre than the decaying gentry in their crumbling houses, isolated and sinisterly eccentric, haunted by the sins of the past" (188).

Yet, the distinction drawn by McCormack is of significant import in the context of nineteenth-century Irish Gothic writing, and in terms of the specific argument of this essay. The extent to which Irish Gothic was, and remained into the late twentieth-century, fixated with the political, economic, and cultural exertions of Ireland's colonial relationship with England provides the motive force, as well as the temperamental anxiety, of an Irish Gothic literature that was principally authored by Protestant, Anglo-Irish writers. As we shall see in more detail below, Irish Gothic is largely founded upon the divisive and violent politics of colonialism. And this is a point that informs Julian Moynahan's survey of Anglo-Irish writing and history from Maria Edgeworth to Elizabeth Bowen and Samuel Beckett, in which he addresses the Gothicization of Ireland within the Anglo-Irish literary tradition. Dealing specifically with Sheridan LeFanu and Charles Maturin, Moynahan summarizes the factors that contributed to Ireland's assumption of a quintessential Gothic geography:

> Politically oppressed, underdeveloped in the far west and southwest, disrupted and distressed by famines, clearances, uprisings, and the depredations of rural secret societies, devoutly Catholic in its majority population, and full of romantic scenery and prehistoric, not to say feudal, ruins, nineteenth-century Ireland was an impressive candidate for Gothic treatment. (111)

While the Irish landscape furnished, as Moynahan outlines, a quintessential Gothic milieu in which narratives of Protestant anxiety could be played out, the histories imprinted onto that landscape, and that, partially, manifested in ruined castles; impoverished native dwellings; and demesned properties, were a source of individual and communal suspicion and fear within the landed Anglo-Irish populace.

Thus, the topographies of Ireland were aligned with fears of native retribution against the regnant Protestant community within the plot arcs and thematic foci of Anglo-Irish Gothic writing across the nineteenth-century. For the Anglo-Irish Ascendancy "the mode was a way of embodying social fear [. . .]. The Gothic encouraged besieged Protestant elite to dramatize its fears and phobias in a climate of inexorable political decline" (Irish Classics 382–83). Decline and anachronism are the temporal signs of the increasing pressure exerted upon the Anglo-Irish, Protestant community in Ireland, and Gothic narratives chart the doomed future and restless past of this constituency. But what is of equal significance when we broach the literary history of the Anglo-Irish Ascendancy in the nineteenth-century, and into the twentieth-century, are the spatial coordinates of that literary tradition, and how these coordinates interplay with the temporal axis of irreversible historical decline. And the most enduring spatial theatre within this literary tradition is that of the "Big House" novel, which, itself, has intimate connections with the broader history of Irish Gothic writing.[2] Within the "Big House" narrative lineage, then, explicit historical decline and geographical isolation engender moods of Gothic paranoia and insecurity. As Vera Kreilkamp argues, drawing the "Big House" novel within the generic scope of the Gothic:

> This turn to the Gothic, apparent in the settings of many national tales and subsequent domestic Big House novels, reflects the fears of an increasingly beleaguered Ascendancy society, trapped in overbuilt but decaying homes, surrounded by a newly resurgent Catholic nationalism, and forced to confront its failure to win native Irish allegiance [. . .] The Irish Gothic novel stylistically and thematically encodes the sublimated anxieties of a colonial class preoccupied with the corrupt source of its power. ("Fiction and Empire" 171)

The "Big House" tradition, in many respects, is an "end of empire" literary phenomenon; the edificial assertion of colonial seizure and power is transformed into a scar on the landscape, a persistent reminder of the usurpations of the past. The "big house" is, then, a moldering embodiment of the unpaid debts of Ireland's colonial history, and the changing political and social temperature of Ireland in the nineteenth-century ominously presage the time when these debts will have to be repaid. It is this sense of the past quaking beneath the very soil of the landed estates, and of the "big house" as a crumbling burden, or as a ruinous prison, that fuels much of the anxiety of the Gothic, "Big House" literature during the nineteenth-century and up until, and after Irish political independence in 1922. As Kreilkamp maintains elsewhere: "Certainly, the political insecurity conveyed by Gothic conventions expresses [. . .] the reality of Anglo-Irish conditions during the long century of local violence accompanying the development of the big house novel" ("The Novel of the big house" 67).

V

Published in his 1933 volume, *The Winding Stair and other Poems*, William Butler Yeats's "Coole Park, 1929," is an encomiastic verse to his friend and mentor, Lady Augusta Gregory. The poem lists, and enlists, many of the creative personalities contributory to Irish cultural nationalist assertions during the Celtic Revival at the end of the nineteenth-century and outset of the twentieth-century, including Douglas Hyde and John Millington Synge. But the poem is equally a paean to a disappearing stratum of Irish society in the wake of Free-State independence in 1922. The titular Coole Park was the ancestral home of Lady Gregory's deceased husband, Sir William Henry Gregory, and was built by his great-grandfather, Robert Gregory in the late eighteenth-century. As with Elizabeth Bowen's *The Last September*, Yeats's verse is a final creative flourish as the energies and the presence of this once dominant social class ebb away. The emphatic nature of Yeats's valedictory verse is revelatory of the obsolescence of the subjects of his plaudits. In Romantic guise, Yeats closes the poem with an acknowledgement of mortality, but also a reminder to future readers to bear in mind the erstwhile genius attached to this location. This final verse situates the "big house" as a venue for, and repository of, all that is to be cherished in fine culture, whereas the index of mortality is figured in terms of a "natural" tide that cannot be stemmed: "Here, traveller, scholar, poet, take your stand / When all those rooms and passages are gone, / When nettles wave upon a shapeless mound" (206). The funereal atmosphere conjured here by Yeats captures equivalent tones and figurations in the late "Big House" narratives explicated below. At a simple level, the levelling of the edificial monuments of feudal colonialism, indeed the real and imagined threats to this class are most often figured in terms of a minatory "natural" hinterland. The land that appeared to be conclusively conquered is imagined, and with good reason in many cases, as restive and rebellious. Decay and cultural decline are, then, slow processes of return to the "natural"—and such processes appear in both figurative and literal terms across the texts under consideration below.

Elizabeth Bowen's *The Last September* (1929) opens with an arrival at the "big house" at Danielstown, the Montmorencys, potent symbols of the "unhousing" of the Anglo-Irish during the early part of the twentieth-century in Ireland, enter the Naylor's estate: "About six o'clock the sound of a motor, collected out of the wide country and narrowed under the trees of the avenue, brought the household out in excitement on to the steps" (7). Yet this auspicious, celebratory and communal beginning is a distraction from the real force of the novel, which is subtly gestured towards in the next paragraph. There may be a dynamism to the first sentence of the novel, but it belies the thread, and threat, of ruination that suffuses much of the remainder of the narrative. Our protagonist, Lois Farquhar, reveals this sense as she watches the guests arrive: "The dogs came pattering out of the hall and stood beside her; above, the vast façade of the house stared coldly over its mounting lawns. She wished she could freeze the moment and keep it always. But as the car approached, as it stopped, she stooped down and patted one of the dogs" (7). While Lois's eye to the future is partly conditioned by the difficult personal choices she faces as a young woman of marriageable age, Bowen's terminology here is suggestive of more grave developments and prospects. The insular social and cultural routines of the Naylors and the Montmorencys across the narrative achieve a "frozen" quality all their own. In fact, the tea parties, dances, visitations, and tennis parties, in the midst of a guerrilla war between the British state and Irish militia, have both tragic and comedic effects. Bowen's lexical selection, "freeze," then, hints at something that is passing, a class, a lifestyle, a set of social mores and habits, that are obsolete, and under threat, from the unseen indigenous terrors that haunt the territories beyond, and within, the demesned estate. And in true "terrorist" fashion, the unseen violences and threats play upon the imaginations of the landed Anglo-Irish; the tension over the material content of the country is enacted equally within the fantasies and darkened imaginaries of the anachronistic "big house" residents.

In a striking passage, again relayed by Lois, and in the wake of a nationalist "outrage," we witness her reaction, as it alters, even enlightens, Lois's perspectives of the actual landscape, and of her situation within the physical and cultural geographies of Ireland. Bowen writes:

> She [Lois] wondered they were not smothered; then wondered still more that they were not afraid [. . .]. The house seemed to be pressing down low in

apprehension, hiding its face, as though it had her vision of where it was. It seemed to gather its trees close in fright and amazement at the wide, light lovely, unloving country, the unwilling bosom whereon it was set [. . .]. And the kitchen smoke, lying over the vague trees doubtfully, seemed the very fume of living. (67)

There is no explicit mention of human agency in this extended rumination; Bowen's anthropomorphizes the Naylor's "big house" in order to underscore the emotional force of Lois's sense of vulnerability and isolation within this physical and cultural topography. In addition, Bowen's deploys a familiar trope within colonial literature as she contrasts the spaces of "big house" civility with the encroaching forest lands surrounding the estate. Of course, the forest forms part of the cosmetic arrangement of the estate, but in Lois's color-schemed view, the competing similes used to figure the house and the forest reveal the relative spaces of security and menace for Lois and her class. The exercises and the burdens of colonial history are pressing hard against the boundaries of the demesne, and they are exerting their force upon the imaginative and figurative faculties of Lois's consciousness. Human presence is implied in this passage in Wordsworthian guise akin to "Tintern Abbey," as we read of the smoke plume above the "big house." In this vein, we might well be reading about a pre-emptive moment of recollection of a past, rather than a description of a "present" experienced by Lois. As Lois travels back to the "big house" as Danielstown after the news of the terrorist attack, her forward motion could easily be taking her into the past rather than towards any viable future at the estate. And this sense of ghostliness in a landscape punctuated with affective and physical wounds of colonial and anti-colonial takes more overt form on another journey made by Lois outside the confines of the "big house."

Together with Hugo Montmorency and Marda Norton, Lois undertakes what begins as leisurely walk along the local river, the Darra. Just as a maudlin Hugo is about to emote to Marda on the condition of his marriage, she interjects with: "Oh, what is that? The ghost of a Palace Hotel?" (122). The walk is interrupted by the prospect of a ruined mill on the banks of the river Darra, and in Bowen's description it achieves the aura of a Gothic artefact: "The mill startled them all, staring, light-eyed, ghoulishly, round a bend in the valley [. . .]. The river darkened and thundered towards the mill-race, light came full on the high façade of decay. Incredible in its loneliness, roofless, floorless, beams of criss-crossing dank interior daylight, the whole place tottered, fit to crash at a breath" (122–23). The moldering ruin's decay and abandonment, its obscurity and its emptiness, are harbingers of the future for the witnesses to its current condition. Within the symbolic economy of Bowen's gothic narrative, the ruination of the mill parallels, and anticipates, the eventual demise through arson of the "Big House" at Danielstown. In-

deed, the vista at the mill also includes vestiges of indigenous Irish ruination
crouching in the shadow of the larger ruin. As the trio progress, tentatively,
they note:

> some roofless cottages nestled under the flank of the mill with sinister pathos.
> A track going up the hill from the gateless gateway perished among the trees
> from disuse. Banal enough in life to have closed this valley to the imagination,
> the dead mill now entered the democracy of ghostliness [. . .] was transfigured
> by some response of the spirit, showing not the decline of its meanness, simply
> decline. (123)

The scene is one of profound Shelley-like levelling; time returns the concreted assertions of hubris, colonial in this case, to fragments. The studied
composure of the architecture of power and conquest is repeatedly revealed
as decayed or decaying in Bowen's text, and across the "big house" genre.
But in this instance there is an explicit alignment between the presence of the
past and the contemporary conditions of violent uprising by the native Irish
populous. Lurking within the frayed structure of the mill, the visitors discover what one can assume is a local Irish "terrorist." As Lois investigates the
interior of the mill with Marda, they disturb the sleeping vagrant: "But the
man rolled over and sat up, still in the calm of sleep. "Stay there," he said,
almost persuasively: a pistol bore the persuasion out [. . .] 'It is time,' he said,
'that yourselves gave up walking. If you have nothing better to do, you had
better keep in the house while y'have it'" (123). Thus we have an anonymous
voice of indigenous resistance, not a voice that is tinged with fear or anger,
but, rather, seems to be assured in how the historical narrative of Irish colonialism will play out. The combination of edificial ruination and the 'terrorist'
are historical legacies, but they are also markers of the future trajectories of
Irish history, along which the physical and cultural integrity of the Anglo-
Irish "big house" society will fracture and collapse. From a postcolonial-
ecocritical perspective, we glean a premonition of the imminent decline of
the anthropocentric composure of the architecture of colonial power in Ireland.

Published in 1970, J. G. Farrell's novel *Troubles* is part of his Empire
Trilogy, which also includes, *The Siege of Krishnapur* (1973) and *The Singapore Grip* (1978). All three narratives explore the decline of British imperial
control across Ireland, India, and Singapore respectively, in the nineteenth
and twentieth centuries. *Troubles* is set during the post-Great War period in
Ireland, which also saw the commencement of the Irish War of Independence
after a decade of sustained cultural, political, and militant nationalist activism. And, though conceived of in the late 1960s, the narrative draws upon
the classical "big house" structure, as it directly confronts the history of
British imperial decline in Ireland, but also, by implication, the emerging
anti-British terrorist war and civil strife in Northern Ireland during the late

1960s. The novel is set in a hotel, The Majestic, rather than a functioning private domestic space, but the residents are of Anglo-Irish stock, and their relationships with the surrounding Irish population are structured by the same rigid class and religious distinctions we note in other "big house" texts. The story relates the experiences of Major Brendan Archer, recently demobilized after a traumatic Great War, at the Majestic in the company of its few residents and proprietors, the Spencer family. In a manner akin to Bowen's *The Last September*, Farrell conveys the increasing obsolescence of the Spencer's and their guests within the local and the national communities. While the main narrative details the eccentricities of life at the Majestic—the defiant delusions of continuing grandeur and relevance—Farrell punctuates the text with extra-diegetic materials that signal the violent and ever-increasing demise of the British Empire in Ireland and across the globe. The core narrative relates the insularity and the whimsy of the Spencer circle, through the eyes of the Major, and Farrell does this in a tone of sympathetic irony. Edward Spencer's indulgences, his prejudices, and his intransigence in the face of impending historical change are related in humorous terms by Farrell, without ever straying into dismissive condescension.

Tellingly, as the novel opens we note the omniscient narration is conducted in the past tense: "In those days the Majestic was still standing in Kilnalough at the very end of a slim peninsula covered with dead pines leaning here and there at odd angles. At that time there were probably yachts there too during the summer since the hotel held a regatta every July" (Farrell 9). The scene is one of both privilege and isolation, and is fitting for the social class, and the Spencer family, at the center of the narrative. The Majestic "had once been a fashionable place. It had once even been considered an honour to be granted accommodation there. [. . .] By the time Edward Spencer bought it on his return from India, however, it retained little or nothing of its former glory" (10). But there is a third temporal perspective present in the opening pages of the novel, the narrator's present. From the previous extract, we note the steady decline of the Majestic, but from the narrator's temporal standpoint, the hotel is now little more than a ruin. Thus, Farrell opens the novel with the revelation that the primary space of the narrative has been destroyed; ruination and absence are thereby signaled as keynotes of the ensuing tale of late imperial decline.

In one of the recurring figurations of the novel, Farrell describes the grand building as not just under threat from external damage, but as being in danger of structural collapse from the interior. In addition to the unseen but very real threats housed in the surrounding geographies, the "big house" faces a "wild" menace within its confines: "The Palm Court proved to be a vast, shadowy cavern in which dusty white chairs stood in silent, empty groups, just visible here and there amid the gloomy foliage [. . .]. The foliage [. . .] was really amazingly thick [. . .] running in profusion over the floor, leaping out to seize any unwary object" (20–21). The shadowiness experienced by the Major in this space is symptomatic of the "insubstantial" nature of the "big house" and its semi-revenant populous. There is something "primitive" to the description of this interior space; the Major seems to be stepping outside the refined geographies of "civility" as he progresses to this recessed location within the Majestic. The vegetation is animated with agency as it gropes and feels for prey from among the static occupants of the Palm Court; in stark contrast to the human population of the Majestic, the vegetation teems with life, its virility is in direct juxtaposition to the sterility and terminal decline of Spencers and the hotel guests. Equally significant is the fact that the plant life flourishing here is typically non-native to Ireland, thus we witness a horticultural correlative to the process of colonization.

As mentioned earlier, Farrell scores his central narrative with extra-diegetic textual insertions, including political speeches and newspaper reports from Ireland and from further reaches of the imperiled British Empire. In a sense, these extra-textual fragments provide necessary historical context for the action of the central narrative; the details of violence, together with the taut political rhetoric reported, impresses the gravity of broader historical events. In the pages, and consciousness of Major Archer, the emerging revolution in Ireland is figured in terms of international revolutionary networks across the globe. Dispatches from Bolshevik Russia and India mingle with reports from local political and militant subversion across Ireland: "Reuters's representative has just had an interview with two Irish girls, the Misses May and Eileen Healy, who have just reached London, having escaped from Kieff with nothing but the clothes [. . .] they were wearing. They tell a terrible story of the Bolshevist outrage [. . .]" (128). While, shortly thereafter, the reader encounters the report of an armed hold-up of ladies and gentlemen en route to a 'big house' ball in Kilkenny (160). At a superficial level, the interpolated extracts recall earlier Gothic conventions, particularly the textual presentation of Bram Stoker's *Dracula*, but in content they reveal the burgeoning threats to civility and to established orders extant in Europe and elsewhere. In another way, Farrell draws the reader's attention to the discursive relationship between his narrative and that of the variegated dispatches, thereby historicizing his own work of fiction and underscoring the "narration" of historical facts by these publications of "record."

Thus Major Archer's consciousness is assailed by the first-hand physical experiences of "big house" Ireland and that of its indigenous hinterland, as well as international reportage. As the narrative progresses the dynamics of the Irish War of Independence begin to press more acutely on the Major, and on his interactions with Edward Spencer. Once more developments are mediated by the press: "At about this time in Dublin a number of statues were blown up at night; eminent British soldiers and statesmen had their feet blown off and their swords buckled. Reading about these 'atrocities' threw

Edward into a violent rage. These were acts of cowardice. Let the Shinners fight openly if they must, man to man!" (384). The attack cited here is, then, both symbolic and physical, but what is equally important is the anonymity of the guerrilla tactics employed by the Irish militant nationalists. This goes against Edward's notions of civil military engagement, and, again, points to the "imaginative" agency of indigenous terror in the face of a much larger occupying colonial authority. At this point Edward is not just referring to the anonymous assaults on the symbolic economy of British colonialism in Ireland, but such sentiments, betray personal anxieties about the potential, even likely, subversion of his own local authority and position.

Yet, if we return to the newspaper reports that deliver national and global incidents to the isolated fastness of the Majestic, there is a growing impression, internationally, that power is slipping from colonial hands, and that, in Ireland, an acceptance of justice and peace is gaining widespread consent. And, tellingly, this is once more figured in cross-colonial terms: "The centre of growing Indian unrest seems to have shifted from the Punjab to the United Provinces [. . .]. Hatred of the landlords is the cause of all the trouble and, undoubtedly, the peasantry has many grievances" (316). The saturation of Irish life by violence is hinted at in the daily experiences at the Majestic, but its national scale is recounted in detail in the press reports read by its residents. Immediately subsequent to the "India" report cited above, we read of a plea for a cessation of such excess: "Ireland is being ground to powder between the two millstones of crime and punishment. For those whose sense of horror recent events have not blunted the daily newspaper has become a nightmare," and this nightmare is caused by the apparent disposability of life in this conflict: "On Monday night a police officer's wife was murdered at Mallow and the officer himself sorely wounded [. . .] in a fight with forces of the Crown, one man was killed and seven wounded" (317). The newspaper reports are, then, part of the imagination of the national community, in Benedict Anderson's terms; they link the imminent decline of the Majestic "big house" to the assaults on colonial authorities across a variety of other local communities in Ireland and, of course, beyond its borders.[3] As we have seen, though the newspapers reports as and are read as "factual," their contents work just as effectively upon the fevered imaginations of besieged colonial authorities.

"History is everywhere," writes Edna O'Brien in her 1994 novel, *The House of Splendid Isolation*, "It seeps into the soil, the sub-soil. Like rain, or hail, or snow, or blood. A house remembers. An outhouse remembers. A people ruminate" (3). The novel is O'Brien's most explicit fictional intervention in the political and religious conflicts that have conditioned much of the history of the island of Ireland across the twentieth century. The narrative explores the intersections of the private and the public under conditions of heightened political and militant pressure. Akin to the texts we have addressed, O'Brien's novel teases out the ways in which national history explodes into the domestic sphere of privacy and routine. Likewise, O'Brien's figuration of History, above, speaks to the gothic imaginaries of the "big house" narratives, which themselves are both haunted and haunting. Combining the critical resources of postcolonialism and ecocriticism, then, re-historicizes experiences of colonial, ecological degradation and exploitation (*Postcolonial Ecologies* 20). In looking at these literary representations of late colonial decline, we get a sense of the residual hubris of colonial authority, but also we can discern the ever-present anxieties engendered by long term economic and tenurial disenfranchisement. The "big house" novel is, as we have seen, part of sub-genre in Irish literary history, with roots in the duration and depths of British feudal colonialism in Ireland. Yet, as the texts under discussion here reveal, "the exterior show of spaciousness and command was intended to mask an inner uncertainty" (*Inventing Ireland* 367). The displays of grandeur and belonging are merely representative of "an attitude, an assumed style" (*Inventing Ireland* 367), which are consistently plagued by, and undercut by, both exterior physical threat and, just as effectively, the gothic evocation of threat within the occupying imagination.

NOTES

1. For a more general application of Marx's thought to ecological imperialism see Brett Clark and John Bellamy Foster, "Marx's Ecology in the 21st Century." *World Review of Political Economy* 1.1, (2010): 142-56. And also John Bellamy Foster, *Marx's Ecology: Materialism and Nature* (New York: Monthly Review Press, 2000), and *Ecology Against Capitalism* (New York: Monthly Review Press, 2002).

2. For an extended treatment of this genre see Kreilkamp's *The Anglo-Irish Novel and the Big House* (Syracuse, NY: Syracuse UP, 1998).

3. Benedict Anderson, *Imagined Communities: Reflections on the Origin and Spread of Nationalism* (London: Verso, 1983).

WORKS CITED

Anderson, Benedict. *Imagined Communities: Reflections on the Origin and Spread of Nationalism*. London: Verso, 1983. Print.

Banville, John. *The Newton Letter*. London: Secker and Warburg, 1982. Print.

Bowen, Elizabeth. *The Last September*. London: Vintage, [1929] 1998. Print.

Clark, Brett, and John Bellamy Foster. "Marx's Ecology in the 21st Century." *World Review of Political Economy* 1.1 (2010): 142–56. Print.

DeLoughrey, Elizabeth, and George B. Handley, eds. *Postcolonial Ecologies: Literatures of the Environment*. New York: Oxford UP, 2011. Print.

Dooley, Terence. *The Decline of the Big House in Ireland: a study of Irish landed families, 1860–1960*. Dublin: Wolfhound Press, 2001. Print.

———. *The Big Houses and Landed Estates of Ireland: a research guide*. Dublin: Four Courts Press, 2007. Print.

Eagleton, Terry. *Heathcliff and the Great Hunger: Studies in Irish Culture*. London: Verso, 1995. Print.

Farrell, J.G. *Troubles*. London: Jonathan Cape, 1970. Print.

Foster, John Bellamy. *Marx's Ecology: Materialism and Nature.* New York: Monthly Review Press, 2000. Print.

———. *Ecology Against Capitalism.* New York: Monthly Review Press, 2002. Print.

Jackson, Alvin. "Ireland, the Union, and the Empire, 1800–1960," *Ireland and the British Empire.* Ed. Kevin Kenny. Oxford: Oxford UP, 2004. 123–53. Print.

Kiberd, Declan. *Irish Classics.* London: Granta, 2000. Print.

———. *Inventing Ireland: The Literature of the Modern Nation.* London: Vintage, 1995. Print.

Kilfeather, Siobhán. "Origins of the Irish Female Gothic." *Bullán: An Irish Studies Journal* 1.2 (1994): 35–45. Print.

Kreilkamp, Vera. "Fiction and Empire: The Irish Novel." *Ireland and the British Empire.* Ed. Kevin Kenny. Oxford, UK: Oxford UP, 2004. 154–81. Print.

———. "The Novel of the big house." *The Cambridge Companion to the Irish Novel.* Ed. John Wilson Foster. Cambridge. UK: Cambridge UP, 2006. 60–77. Print.

———. *The Anglo-Irish Novel and the Big House.* Syracuse, NY: Syracuse UP, 1998. Print.

McCormack, W.J. "Irish Gothic and After." *Field Day: Anthology of Irish Writing.* Vol. II. Ed. Seamus Deane. Derry: Field Day, 1991. 831–54. Print.

Moynahan, Julian. *Anglo-Irish: The Literary Imagination in a Hyphenated Culture.* Princeton, NJ: Princeton UP, 1995. Print.

Mukherjee, Pablo. *Postcolonial Environments: Nature, Culture and the Contemporary Indian Novel in English.* Basingstoke. UK: Palgrave, 2010. Print.

O'Brien, Edna. *The House of Splendid Isolation.* London: Phoenix, 1994. Print.

Ohlmeyer, Jane. "A Laboratory for Empire? Early Modern Ireland and English Imperialism," *Ireland and the British Empire.* Ed. Kevin Kenny. Oxford: Oxford UP, 2004. 26–60. Print.

Slater, Eamonn, and Terence McDonough. "Marx on nineteenth-century Ireland: analyzing colonialism as a dynamic social process." *Irish Historical Studies* XXXVI.142 (2008): 153–72. Print.

———. "Bulwark of Landlordism and Capitalism: The Dynamics of Feudalism in Nineteenth Century Ireland." *Research in Political Economy* 14 (1994): 63–118. Print.

Williams, Raymond. *Culture and Materialism.* London and New York: Verso. 2005. Print.

Yeats, William Butler. *The Collected Poems of W.B. Yeats.* Ware, Hertfordshire, UK: Wordsworth Editions, 2000. Print.

Chapter Eleven

Landscape and Animal Tragedy in Nsahlai Nsambu Athanasius's *The Buffalo Rider*: Ecocritical Perspectives, the Cameroon Experiment

Augustine Nchoujie

In their attempts to explore the field's richness and variety, critics of Cameroonian literature have applied a plethora of Western-inspired literary theories from formalism and structuralism through Marxism and psychoanalysis to feminism and cultural materialism. Many Cameroonian scholars are excited by this seemingly limitless critical pluralism. Yet others claim that none of these exogenous theories can be said to be truly relevant to the study of Cameroonian literature. There is no consensus of opinion about the ideal critical approach to Cameroonian literature since this literature grows in all directions. One of the directions has been inspired by Nsahlai Nsambu Athanasius in the 2008 novel *The Buffalo Rider*, where he attempts to highlight environmental concerns with focus on habitat degradation and the rapid disappearance of various animal species. The environmental perspective expressed in this work provides a basis for prospective ecocritics to experiment with environmentally inflected approaches to reading.

In this chapter, I seek to analyze, by way of this recent novel, the agonies faced by animals in Cameroon, especially the legendary buffalo, and the degradation of the physical environment at the hands of humans. At a time when the world's physical environment and its occupants are threatened, animals remain the most vulnerable in the natural chain—maltreated, discriminated against, sold, butchered, eaten, and abused as objects of no consequence. *The Buffalo Rider* defines the grounds upon which the human and the animal realms can coexist and flourish in the biosphere. Unfortunately, the animals and the physical landscape frequently become victims of human insensitivity—this is true in West Africa and in many other parts of the world. By fictionalizing this human insensitivity, Athanasius seeks to raise awareness about the daily agonies and abuses these animals face across the globe and for Cameroonian people in particular and Africans in general. He shows how local species of animals form part of the eco-chain, part of God's natural weave, and should not be marginalized and exploited.

LOCATION AND BRIEF HISTORY OF CAMEROON

Cameroon, which is situated very close to the heart of Africa, is bordered in the West by Africa's most populous country, Nigeria, and in the north by Chad; it is flanked on the East by the Central African Republic and shares Southern borders with Equatorial Guinea, Gabon, and Congo. Cameroon's coastline lies on the Bight of Bonny, which is part of the Gulf of Guinea and the Atlantic Ocean. It is a national triangle peopled by more than two hundred ethnic groups and with as many languages as there are ethnic groups. This has earned Cameroon the appellation "Africa's Linguistic Tower of Babel." Culturally and geographically, the country is known by many as "Africa in miniature," while politicians fondly refer to it as the "Bread Basket of Africa" because of the abundance and diverse kinds of food that Cameroon produces. It is also considered as one of the most "peaceful nations" in a turbulent region. Cameroon is a special nation with a unique history.

It is one of the countries in Africa that has had more than two colonial masters. In 1884, during the infamous Berlin Conference, Cameroon was taken over by the Germans as a protected territory. It became a German protectorate from 1884 to1916, at which time the Germans were ousted by the combined forces of the British and the French during World War I. The League of Nations later repartitioned Cameroon, following which the country became a UN Trustee in 1946. The French were given the greater part of the country, while the English received a smaller portion known then as the Southern Cameroons. The Southern Cameroons was administered as part of Nigeria. In January 1960, French Cameroon gained independence, while the Southern Cameroons, in a UN organized Plebiscite on the October 1, 1961, decided to join French Cameroon, forming the Federal Republic of Cameroon. As a result of this colonial history, there are now two principal cultures in Cameroon—the French and the English.

The Francophones have dominated the world of letters, whereas it was a slower literary growth trajectory for their Anglophone counterparts who were partly imprisoned by a dictatorship that shunned letters, especially from a people who were marginalized in all spheres of public life. The lack of publishing houses also proved to be a major deterrent. It was only in 1993 that the first workshop on Anglophone Cameroonian literature was held at the German-sponsored Goethe Institute in Yaounde. The silence of Anglophone writing was broken in the 1980s and 1990s, which experienced an upsurge of Cameroonian Anglophone literature with writers like Bole Butake, Bate Besong, Linus T. Asong, Peter Nsanda Eba, Azanwi Nchami, Francis Nyamnjoh, and a host of others.

Corruption rocked all facets of the nation for four decades—dictatorship, favoritism, marginalization of Anglophones, and wanton destruction of Anglo-Saxon values. Ecological issues were completely shelved and pushed into oblivion. Most of the literary works from both sides of the Mungo (the River Mungo is the historic river that demarcates the boundary between the French-speaking and English-speaking Cameroons) have been more concerned with the socio-political, economic, and cultural issues than with the environment. Problems like the rapid advance of the Sahara Desert, deforestation, logging, poaching (the infamous "bush meat" phenomenon), and now the much-talked-about climate change remained peripheral in the major literary texts produced in Cameroon as most of the writers from both sides of the divide remained timid in their defense of the environment, or perhaps distracted by social concerns. The environment has been used in most cases simply as a setting with little focus on the health and fate of the land.

THE AUTHOR

According to biographical notes, Nsahlai Nsambu Athanasius was born on the August 3, 1954, in Nso, a member of one of the main tribes in the Northwest province of Cameroon. He was educated at the Presbyterian schools in Vekovi, Nso, Dikume, and Ndian in the Southwest province of Cameroon. He proceeded to Cameroon Protestant College, Bali; the University of Yaounde; World Maritime University, Sweden; and the International Maritime Academy-Italy, obtaining a B.A degree in English and French and an M.Sc in Maritime Administration. His bilingual degree propelled him into administrative roles. He was eventually named Chief of Service for Limbe Maritime District, Charge d'Etudes Assistant at the Ministry of Transport, and Sub Director of Maritime Transport at the Merchant Shipping Department in Douala. Currently, he is the Technical Adviser at the Port Authority of Doula-Cameroon. As a maritime administrator, he has written "Questions and Answers on the Safety of Small Sea-Going Craft." His love for letters eventually found expression in his first novel, *Out of the Shadows*, a 2004 publication which examines the dynamics of Nso culture. *The Buffalo Rider* is his second artistic endeavor and ventures into an area of national and international concern—the fate of animals in a threatened biosphere.

ANIMALS IN THE CAMEROONIAN SOCIO-CULTURAL AND
LITERARY IMAGINARY

Cameroon is well known for its great human-animal-nature tradition. Animals were and are part of the people's cosmology and belief system. In Cameroonian cultures, animals occupy a significant position, especially in the socio-cultural realm. As both symbols and subjects, they are not only considered a part of the cosmic creation, but are associated with a spectrum of environmental issues ranging from bioregionalism, heritage, identity, and environmental justice, among others. They play major roles in various rituals. Domestic animals were considered as "relatives" and "neighbors," not "others" living in another distant world severed from the human world, not pets to be turned to for companionship in times of agony and distress or a therapy for boredom, loneliness, and insecurity. They are family members whose names were reminiscent of family values, dreams, and aspirations. The West African mind, indeed the human atmosphere, is, to echo Paul Shepard in *The Others: How Animals Made Us Human,* "the result of a long series of interactions with other animals" (15). Though there is a kind of hierarchy in the Cameroonian relationship to animals, it is a hierarchy built on reciprocity, a reciprocity that sought to mitigate excessive anthropocentric tendencies in the relationship. This harmonious chain was rudely disrupted with the encroachment of colonialism and its overwhelming impact on the colonized peoples of the world, including Cameroon. Even animals were not spared. Graham Huggan and Helen Tiffin in *Postcolonial Ecocritcism* corroborate this when they state:

> Aware of the power of animal spirits, native hunters treated their prey with respect and performed rituals defined by reciprocity. Although not quite a relationship of equals, the connection between Indians and prey was not essentially hierarchical. But notions of domination and subordination were central to the English, who believed that the act of hunting epitomized the divinely sanctioned ascendancy of humankind over animals. (10)

In Africa, animals and humans lived in relative peace until this harmony was disrupted by colonial influences. In Cameroon, animals suffered the same fate as their counterparts elsewhere in the colonized world. Athanasius's *The Buffalo Rider* highlights some of the disturbing ruptures of the entente between animals and humans, a situation that has radically transformed animals into "others," a dangerous relationship Paul Shepard decries when he laments that animals are "others in a world where otherness of all kinds is in danger, and in which otherness is essential to the discovery of the true self" (5). For the purpose of this chapter, my focus will be on the buffalo, which is considered one of the "big five" in the African jungle.

The Buffalo Rider pays tribute to one of the greatest animals in the African jungle. Considered as one of the "big five," the African buffalo is a symbol of courage, power, force, strength, and endurance. The hunting of this animal has traditionally been reserved for royalty. No commoner could go hunting for a buffalo without the royal blessing. In the Northwest and Western provinces of Cameroon, the buffalo was held in particularly high esteem, respected and revered for its royal trappings. Some of the greatest artefacts from the Northwest and Western provinces, especially priceless stools that are showcased in international museum tours, have paintings of the buffalo on them. Such stools were only made for members of the royal family and for guests that deserved the highest honor of the land according to the reckoning of the Chief, the Fon, or the Sultan. Because of its fierce nature, the buffalo was considered dangerous and nicknamed "the widow maker" or "Black death."

But in the novel Athanasius paints a rather unusual picture of the buffalo. In Nsoland, it is the dream of every hunter to hunt, kill, and bring home a buffalo to the palace in order to be recognized with the cherished title of "Fonkwah," which means "great hunter." It is a symbol of power and pride. Whoever accomplishes such a feat is rewarded by the Fon and is held in high esteem by the royalty. Their social status gains considerable significance in the community. Their caps are adorned with bright red feathers—a passport to royal gatherings and all "Manjong houses" in the land. Manjong houses are all-male groupings whose mission is to ensure social cohesion in the village and the respect of the laws of the land. The killing of the buffalo, therefore, is seen as a profoundly communal event. However, with this great premium placed on dead animals in the name of titles and royal recognition, the buffalo and other animal species became endangered.

Generally speaking, Africans, including the Nso people featured in Athanasius's novel, respect animals and have a biophiliac relationship between themselves and other natural forms. Consequently, a number of indigenous animal preservation practices were en vogue before the advent of colonialism. Though hunting was a major activity for the menfolk, it was greatly regulated and controlled by traditional norms. There were particular seasons and areas where hunting could take place. It was a taboo to kill an animal that was "pregnant" or "nursing." Animals caught mating were also to be spared. If any of these rules were violated, rituals were performed. Indigenous taboos and superstitious beliefs consciously acted as a deterrent against animal slaughter by title seekers in search of fame. Even animals like the tortoise, because of its perceived wisdom, have omens attached to their killing. In Nsahlai's novel, the character Baa Tanle expresses fear when he sees a hunter carrying a tortoise: "I saw one of the hunters carrying a big tortoise. . . . I wonder whether you know that a tortoise is a beast of bad omen" (129). If these omens were anything to go by, then they would act as a means of preserving some of those animals that are seriously endangered. The author of *The Buffalo Rider*, an emerging voice in the Anglophone literary landscape in Cameroon, defies the norm by creating awareness about the fate of the environment and the non-human subjects inhabiting the landscape. The novel describes with vividness and simplicity the buffalo's search for its land of birth, a passionate journey rendered gruesome by human obstacles, the loss of its once glorious trails now transformed into massive farmlands and human habitats. The buffalo, now a stranger in its once safe haven, is no longer at ease. With humans chasing it, it carries Baa Tanle's on its back like a tick and ferries him to Mbar, before it is eventually slaughtered in cold blood in the name of honor and fame. It is worth mentioning that Baa Tanle is one of the most experienced hunters in the village. A skilled horse trainer, he is noted for his equestrian antics. He is so skilled that the villagers nickname him "ngimri" which means the tick, for he has the ability to gum himself on a horse like a tick and no amount of running, jumping, and kicking will dislodge him. He is simply the best hunter in the land. With the arrival of the legendary buffalo, he feels threatened in his position as the best hunter in the land. In order to maintain the status quo, he butchers the buffalo at its most vulnerable moment. This happens, tragically, at a time when the animal feels it has Baa Tanle for a protégé in the midst of human aggression.

It is because of the accent on the depletion of the environment and the tragedy that befalls the buffalo and other animals that I single out *The Buffalo Rider* as a "Grande premiere" for Cameroonian literature in general and Cameroonian Anglophone writing in particular. The issues in the text are relevant today across the globe: the insensitive and continuous exploitation of the environment by humans and the fate of animals in an increasingly unfriendly and ravenously meat-consuming world, which is impervious to animal wellbeing and indigenous ecological wisdom.

The story of the buffalo rider bears witness to the fact that African and Cameroonian culture is interested in understanding animal consciousness, communication, and emotions. The perilous journey Baa Tanle and the buffalo embark on testifies to this. The relationship between Kpulaban and Baa Tanle is more than a testimony of how Africans can interpret the behavior of their animals. The encounter between Pa Kintang and Tav Kobang (in the struggle to destroy Pa Kintang's crops) and the chimpanzees also shows that animals have emotions just like us, that it is important to understand the integrity of all ecosystems and to preserve biodiversity and the resilience of life-supporting systems.

The Buffalo Rider depicts the agonizing and epic experience of a buffalo whose attempts to "trace its roots" meet with human and non-human obstacles. The animal highway has been invaded, and the beasts are left in a wilderness of want. Despite the buffalo's prowess in the face of adversity, it is cruelly butchered by its human companion, raising a crucial issue in ecocritical discourse—the fate and place of animals on the ecosphere. In narrating this seemingly epic journey, this "animal tragedy," Athanasius paints the once glorious landscape—the flora and fauna in all its cosmic beauty—as a realm now bare and barren, soliciting a "reromanticization" of the environment and making a strong case for the defense of the environment and its nonhuman subjects from human cruelty and eventual extinction. After the horrendous "murder" of the mythical buffalo, Athanasius laments with rhetorical flourish, "Man The Killer has done it again! Cry the beloved animal kingdom. Cry the defenceless creatures. Who shall help the helpless lot?"(133).

Landscape is an important semiotic feature in ecocritical discourse. It is a geographical territory, a milieu, a physical field of significance with a network of meanings for its occupants, a place that the inhabitants do not simply occupy but experience it in deeply psychological and even spiritual, ways. Simply put, Leslie Marmon Silko in "Landscape, History, and the Pueblo Imagination" defines landscape as "a position of territory the eye can comprehend in a single view" (264). Likewise, in Cameroon, landscape is an integral component of human existence. It has physical, social, psychological, and religious ramifications for the people who live on it. It is an inseparable part of the people's view of the world. It is the people's wellbeing. Consequently, the Nso people, and Cameroonians in general for that matter, have traditionally revered habitat and valued it more than any of their possessions.

Athanasius, in *The Buffalo Rider*, has made an attempt to paint the landscape holistically with particular focus on the mountains, the hills, rivers, valleys, plants, and animal species. The story takes place in Nsoland with the Mbar-Ngongmbaa-Kilum Mountain chain as a prominent scenario for the unraveling of the buffalo tragedy:

> It is conveniently implanted into the South-eastern skies of the Nso Fondom where it wakes the sun up from its nightly slumber . . . . To watch Mount Mbar and its clothes of trees and grasses takes on indescribable hues is to participate in a feast of cosmic beauty. (3)

Athanasius's attempt at using personification to delineate the landscape is appealing, and the narrative is replete with the images of Mbar Mountain which "wakes the sun up from its nightly slumber . . . , while its "clothes of trees and grasses take on indescribable hues" (3). Far from being an exaggeration, participating in a "feast of cosmic beauty" and watching the mountains are ecocritically friendly. Nature is seemingly given its grandeur, even in very humble ways. John Tallmadge in "John Muir and the Poetics of Natural Conversion" posits that, "personification is the highest form of flattery, bespeaking reverence for nature and biotic egalitarianism" (73), and we see something similar happening in Athanasius's representation of the Cameroonian mountains. Along this mountainous stretch, there are the Ngongmbaa highlands and the Kilum Mountain, each like a rocket shooting high up into the Western sky. It is this mountain range that becomes a magnetic force, pulling the buffalo to its once beautiful mountain with all its natural forces intact. The reader is informed that "the foot of Gwan Mbar, with its many streams, rivers and lakes encompasses vast plains and swamps" (4). Mbar is a veritable "nature" sanctuary, a reserve where streams, rivers, lakes, plains, swamps, grasses, animals, and salty springs still enjoy considerable freedom and abundance. Unfortunately, humans in their anthropocentric arrogance, lord it over the "others." Humans disregard other natural forms. They launch an onslaught against these "natural forms" in order to ensure and maintain their supposedly given, biblical dominion. Nature lies helpless, begging to be given a place in the biosphere. Athanasius laments:

> Today Mbar looks barren, ugly, and sad. Its forests are dwindling, its wildlife is lamentably diminishing, its grasses are more stunted and withered and its beauty is fast fading as fame seekers, dealers in animal and animal parts, and gluttons continue to reign havoc on its richness. (133)

Man's intensive farming methods, use of sophisticated weapons of mass animal destruction, tree felling, unorthodox hunting practices have all helped to render Mbar, the reserve, barren. This is Athanasius's fictional portrait of the degradation that the environment has experienced in Nsoland. In reality, trees and forests are disappearing in Cameroon. The Sahara Desert is advancing and almost engulfing the Northern provinces. This has forced the Cameroon government to start off the "Operation Green Sahel" with intensive tree-planting campaigns across the Northern provinces.

According to Lawrence Buell, one of the hallmarks of an environmentally oriented work is that "the human interest is not understood to be the only legitimate interest" (7). This is certainly true in Athanasius's *The Buffalo Rider*, for the non-human interests seem to override the human interests. Here the accent is on animals. The place of animals in the ecosystem has become a subject of great debate in recent ecocritical discourse. This has always been a controversial area among ethicists and ecologically oriented intellectuals and scientists. Today more than ever before, ecocritics, environmentalists, deep ecologists, and ecofeminists are calling for a more humane approach and attitude towards animals, which form an integral part of the biotic community. Animals are frequently seen as providing lessons to a debauched technologically blinded and psychologically deficient human community, which for centuries has taken pride in a certain exclusive humanistic arrogance. Jody Emel and Jennifer Wolch allude to this all-important role of animals in the following words:

Animals have been so indispensable to the structure of human affairs and so tied up with our visions of progress and the good life that we have been unable to (even try to) fully see them. Their very centrality prompted us to simply look away and to ignore their fates. But human practices now threaten the animal world and the entire global environment as never before. Our own futures are on the line too. Hence, we have an intellectual responsibility as well as an ethical duty to consider the lives of animals closely. (*Animal Geographies* xi)

The above view is corroborated by Steve Baker, in *Picturing the Beast: Animals, Identity, and Representation*, who contends that:

Much of our understanding of human identity and our thinking about the living animals reflects—and may even be the rather direct result of—the diverse uses to which the concept of the animal is put in popular culture, regardless of how bizarre or banal some of those uses may seem . . . . Culture shapes our reading of animals just as much as animals shape our reading of culture. (4)

But over the centuries, man has systematically ignored these very important "neighbors" who also deserve consideration and a place on the ecological map of the world. Today, ecocritics and environmentalists are saying that time has come for humans to recognize that they are not the only occupants of the earth and that their clinging to their supposed supremacy and uniqueness on the grounds of reason vis-à-vis the animals is becoming naively anthropomorphic since animal instincts and human reason are merely different degrees of the same quality. Mary Midgley postulates:

The tremendous concentration of interest on human intellectual achievements is still so central to our thinking that we have difficulty in criticising it at all, let alone in doing so fairly. But in the last half-century, its shortcomings have grown increasingly glaring. Of course what is good about humanity must be celebrated and cherished. But the attempt to do this by sacrificing everything else—by cutting humanity off from its context, by rejecting everything non-human and non-rational as valueless—no longer looks today even like a rational project, let alone a realistic one. This change in our attitude is still far from complete. (32)

There is a dire need for this cycle to be complete, for man to relinquish some of his "humanistic arrogance" and come to terms with the reality that he is no longer the only creature that inhabits the earth and that his continuous attempt to ignore the other actors in the biosphere will endanger man and the entire world.

In *The Buffalo Rider*, a miniature kingdom of animals is captured, from birds to baboons, gorillas, monkeys, buffaloes, frogs, lizards, tortoise, and lions. The buffalo and its rider are competing for the protagonist position in the novel. There is the buffalo and Baa Tanle on the race for heroism—

human or nonhuman. The buffalo sets out for its historic journey to revisit its land of birth as it snorts, then sniffs, lifts its head high and starts off for the Mbar Mountain, which stands tall and mesmerizing. The buffalo visualizes the appetizing greenery of Mbar and smells the sweet aroma from its multiple flowers. All these fill the buffalo with nostalgia. Oblivious of any obstacles that might come its way the buffalo sets off at a trot, then a canter and a gallop toward the inviting land of its birth. It is during this journey that the buffalo discovers with dismay that the old animal highway, the trail or track it had followed, has been invaded by man. The buffalo trots on through numerous valleys, over undulating hills, under grooves, through thickets, and across ravines. At Vekovi, it meets with stiff human resistance as the residents team up to put an end to its advance. Many bullets are fired at very close range but the buffalo defies all; even Paa Tangwah's high-precision, deadly, double-barreled gun, which is known to have terminated elephants, leopards, hyenas, and baboons, is snubbed by the enraged buffalo.

News of the deadly and resistant buffalo has reached all the nooks and crannies of Nsoland as the buffalo, now suffering with bullet wounds, continues its advance with elephantine fury. At Jakiri, all attempts to neutralize it fail and Baa Tanle, an all-time expert hunter and last hope for the people, is sent "flying like a balloon into the air" by the buffalo. Up in the air, "Baa Tanle's spears had abandoned him and fallen to mother earth as if to go and report to her what had befallen their owner and master" (25). Instead of falling on the deadly horns, he lands on the hump like any cat would and sticks to it like gum. For three days and nights, Baa Tanle is stuck to the back of the buffalo en route to an uncertain destination. It is through this perilous adventure that the animal kingdom is exposed, the relationship between animals and animals and between animals and man examined.

When Baa Tanle carries out his seemingly heroic act of killing the buffalo that he has befriended for three gruesome days, Taa Ngitir is so eager to know from Baa Tanle about the animals and the weather at the summit of the mountain. In response to this enquiry, he says: "All I can tell you about animals and Mbar is that Mbar is an animal kingdom" (128). Mbar is a veritable jungle where 'jungle survival' reigns supreme. In one of the valleys along the Mbar-Ngongmbaa stretch, Baa Tanle and the buffalo run into a wild-monkeys' circus with a troupe of monkeys displaying their acrobatic skills from the heights of swaying tree-tops. Hundreds of lizards could be seen crawling playfully on backs of trees and on rocks. Anthills are pictured searching for insects to prey on for food. Cattle egrets are seen feeding on insects. At Mbokam Hunting Bush, birds of all kinds and animals descend on a giant antelope struggling to free itself from a man's trap. Before Baa Tanle and his companion can arrive, hounds tear off pounds of flesh from the helpless antelope. Humans would certainly look to this and say, "Yes, they behave true-to-type! Survival of the fittest!" These are isolated cases of disharmony in the animal community. If humans, who lay claim to a superior intelligence, reason, and a moral code, could go into the laboratories to painstakingly fabricate chemical weapons to kill fellow men, to fabricate deadly inter-continental ballistic missiles and B52 bombers that could destroy thousands of humans in a minute, are the animals not better in their instinctive search for food in order to survive? Humans must therefore remember that animal instinct and human reason are forms of intelligence. Animal instincts should not be minimized and considered inferior by virtue of their "animality." After all, as Greg Garrard in *Ecocriticism* (2012 edition) quotes Timothy Morton as saying, "humans are like 'animals,' but 'animals' are not 'animals'" (160).

Despite their predatory instincts, the animals tend to become united when their kind is in danger. In one of the spots along the animal highway, Baa Tanle and the buffalo encounter gorillas. The gorillas "were visibly angry" when they saw a man stuck to the back of a buffalo. They are frightened but determined to wage a war of liberation. Athanasius writes: "They groaned and growled and barked like a possessed bundle of infuriation; they rushed around collecting fallen tree branches and stones with which to undo the insult on the buffalo's back" (92). This is a mark of camaraderie among the animals to chase away the unwanted rider of their fellow animal. In another instance, the two travelers run into a huge colony of birds. The birds are disturbed to see a man sitting on the back of a buffalo, one of the kings of the forest. To the birds, this is an aberration, an awful sight, and they call for a massive and vehement protest. The birds and the gorillas want their "brother" freed from the human chains. These protests are signals to man that animals should be treated equally or even just "humanely." They also deserve a place in the ecosphere. Some animal activists even think that the animals should not just be treated "humanely" but should be given equal rights, legal rights. From the above protests, it is clear that the relationship between man and wild animals is built on confrontation. A more dramatic example is the relationship between Baa Tanle and the buffalo. After going through valleys and thickets, crossing dangerous rivers and ravines, and climbing hills, Baa Tanle is suddenly taken hostage by the idea of fame despite deep meditations on the consequences of such an act. He contemplates: "I will not die without a name. A real name! I will not die without a tall red feather in my cap. I am at the threshold of honour and pride. I must bring honor to my Fon, the Nginyam of Nso. I must bring honour to my family" (107). Paradoxically, this honor is to be achieved through a great act of dishonor to the buffalo.

He eventually decides that he is going to kill the buffalo whose back he has clung to for three days. The animal is physically exhausted and at its most vulnerable moment, Baa Tanle decides to carry out his treacherous act, depriving the animal of its life and the heroic mission it had set out to

accomplish—to visit the land of its birth in order to worship its once flourishing landscape. To carry out such an act, Baa Tanle has to summon courage:

> Courage, he now understood, was not the total absence but the complete con-
> quest of fear. His body stiffened. He sweated profusely and rivers of sweat
> coursed down his body. His right hand was holding the thirsty knife firmly so
> that the muscles on it stood out under tension. He breathed in, held his breath
> and closed his eyes. And then it happened! (117)

The buffalo is cruelly butchered. Baa Tanle's behavior before the act is
indicative of the moral implications involved in this act of cruelty. Jacques
Derrida, in an article entitled "The Animal that Therefore I am (More to
Follow)," does not hesitate to compare this kind of treatment of animals by
humans to "The worst cases of genocide" (20). There is immense suffering
captured in this act of killing.

The buffalo has not only suffered the pangs of the sojourn on an en-
croached pathway, but also the excruciating pains of the deadly knife. This
indeed is a great tragedy, not only because the buffalo is murdered, but also
because of what happens after this incident. The killing of the buffalo acts as
a motivating factor to other title seekers to screw up their courage and go on
an all-out hunt for more animals to kill and bring home in order to be
honored as heroes. In the Epilogue of the text, Athanasius explains how this
unprecedented crave for titles and honour galvanizes the village folks who
promise to outsmart Baa Tanle by bringing home alive the most dreaded
beasts of the land such as leopards, hyenas, elephants and even lions!

This, in a nutshell, is Athanasius's fictionalization of the daily agonies of
thousands of animals across the globe who are crying out for mercy from
their human counterparts, but the cries seem to go unheeded as countless
animals remain objects of varied abuses and undeserved occupants of slaughter houses across the world. This indeed is tragic. Despite this grim picture,
Athanasius has demonstrated that animals are still capable of a livelihood
and a personality, that they possess a certain degree of intelligence and are
active agents with intrinsic value.

As a young hunter, Baa Tanle has been warned against chimpanzees for
they are not only physically strong but are also intelligent, capable of imitat-
ing men and distinguishing men from women they would threaten and fright-
en women if they find them alone in the farms. This is what happens to Pa
Kintang when he takes ill during the harvest season. The chimpanzees take
advantage of his absence, frighten his wife and daughter and ravage the
crops. They are only caught when Pa Kintang and his friend, disguised as
women and armed with deadly weapons, attack, kill, and chase away the
chimpanzees. Charles Darwin in *The Descent of Man*, in explicating natural
selection, argued that

The difference between the mind of the lowest man and that of the highest animal is immense but the differences were of degree, not kind: "The senses and intuitions, the various emotions and faculties, such as love, memory, attention, curiosity, imitation, reason etc, of which man boasts may be found in an incipient, or even sometimes in a well-developed condition in the lower animal." (104–05)

The chimpanzees make gender-based decisions: women are weaker, they can be frightened. They experiment and it works out. This is a signal that animals are endowed with a humanlike intelligence. Like the chimpanzees, the buffalo remains one of the greatest cultural signifiers in the land. Despite its importance in "making humans more human," as seen in the ways its parts are used, humans still rein havoc on it, killing it in order to attain human grandeur. The cruel slaughter of the buffalo, despite the tremendous effort it has exhibited, is testimony of the anthropocentric view of the animal world—animals are objects to be observed and used. It is in this connection that Greg Garrard quotes John Berger as saying: "Animals are always observed. The fact that they can observe us has lost all significance. They are the objects of our ever extending knowledge. What we know about them is an index of our power and thus an index of what separates us from them. The more we know, the further away they are" (Berger 14). This is echoed by Arran Stibbe in an article titled, "Animals Erased: Discourse, Ecology and Reconnection with the Natural World." He makes reference to one of the most scorned animals—the pig—wondering aloud what would become of such animals which for centuries have remained objects, treated with disdain. One is tempted to wonder aloud why Baa Tanle would, in cold blood, butcher the buffalo, an animal that is revered in the land, a three-day companion, for a title. This is the objectification of animals that has gone on for long. By dramatizing the buffalo-Baa Tanle encounter, the author calls to question the role of animals in African cosmology, igniting a forceful debate on animals as cultural signifiers, raising a series of hair-splitting questions: Should we kill animals in order to sustain our human enterprise? Should we eat them as food in order to survive while they perish in our stomachs? How can a more humane and harmonious connection be made with these creatures, who by no fault of theirs, have remained appendages in the human realm? As Cameroon's "Grande premiere" of Anglophone fiction, *The Buffalo Rider* is an eye-opener to the intrigues and intricacies that surround ecological discourse in Cameroon, the rest of Africa, and beyond. This is certainly the beginning of the de-objectification of animals in Africa and African literature, and the ecocritical approach enables readers to focus on the crucial relationships between humans and other species that are highlighted in the literary text. African indigenous wisdom views animals as God's creatures who serve a place in the cosmos. The ecological implications of the obliteration of animals throws

powerful moral urgency into our understanding that it is necessary to recon-
nect with the reality of animals if we are to build human systems that work
with, rather than against, nature.

At this juncture, I would contend that Athanasius's literary craftsmanship
is a step in the right direction. It attempts to redirect literature towards animal
and environmental protection. Cameroon's animals, indeed many African
animal species, are endangered. Humans have invaded their animal world
and driven them out of their habitats. There is no hiding place for animals in
Africa, which once was an animal citadel. Population explosion and excessive animal consumption have driven many animals to extinction and are
imperiling other species. Athanasius is one of the literary voices crying out in
the wilderness for Cameroon to protect its animals and plant species. His is a
call for a return to the primary indigenous ecological strategies that were
used in Africa, notably the belief in the divine nature of animals. As divine
beings, some of them, especially those endangered, had taboos attached to
their killing. This deterred reckless animal slaughter. Unfortunately, most of
the taboos have been demystified by modernism.

I find myself wondering if the belief in omens and fear of consequences
of animal maltreatment should be revived, if it *could* possibly be revived by
way of literature. This clarion call seems to have made a difference as more
and more individual Cameroonians, organizations, and governments are becoming more aware of the loss of rich biodiversity of Africa. Thanks to the
creation of Cameroon's Ministry of Environment, Nature Protection, and
Sustainable Development in 2004, a good number of national parks, notably
the Korup National Park and The Waza National Park, are bringing about
marked changes in the preservation of wildlife. To demonstrate the government's commitment to protecting endangered species, the Cameroon government fought for close to six years to bring home Cameroonian gorillas that
were stolen, smuggled from Cameroon to Nigeria, South Africa, and Malaysia in 2002 and kept illegally at the Taiping Zoo. They were eventually sent
back to Pretoria. After years of diplomatic wrangling and effort to determine
the nationality of the gorillas, a DNA test was conducted and it was con-
firmed that the gorillas were Cameroonian. They became known as the
"Taiping Four." It was a great diplomatic victory for Cameroon as a twenty-
four-man convoy ferried the "Taiping Four" gorillas to Cameroon amidst
fanfare. That was on November 29, 2007. The triumphant entry and reception of the gorillas in Cameroon is the kind of "practical ecocriticism" that
we look forward to in Africa: a revisiting of all indigenous ecological preservation practices and knowledge, effectively and meaningfully combined with
modern methods, not only theorizing ecocriticism in books but also engaging
in meaningful praxis. There is no doubt that this combination might just be
the much needed ecological therapy that will save the world whose fate is

boldly written on the wall of doom if humans continue to turn a blind eye every day to countless warning signs.

WORKS CITED

Athanasius, Nsahlai Nsambu. *The Buffalo Rider*. Douala, Cameroon: NNAMB'S Publishers, 2008. Print.

Baker, Steve. Picturing the Beast: Animals, Identity and Representation. Champaign-Urbana: U of Illinois P, 1993. Print.

Berger, John. "Why look at Animals?" *About Looking*. London: Penguin, 1980. Print.

Buell, Lawrence. *The Environmental Imagination*. Cambridge: Harvard UP, 1995. Print.

Cooper, David E., and Joy A. Palmer, eds. *The Environment in Question: Ethics and Global Issues*. London, UK: Routledge, 1992. Print.

Darwin, Charles. *The Descent of Man*. 1871. Princeton, NJ: Princeton UP, 1981. Print.

Derrida, Jacques. "The Animal That Therefore I am (More to Follow)." *Critical Inquiry* 28.2 (Winter 2002): 369–419. Print.

Emel, Jody, and Jennifer Wolch, eds. *Animal Geographies: Place, Politics, and Identity in the Nature-Culture Borderlands*. New York: Verso, 1998. Print.

Friday, Laurie, and Ronald Laskey, eds. *The Fragile Environment*. Cambridge, UK: Cambridge UP, 1989. Print.

Garrard, Greg. *Ecocriticism*. New York: Routledge, 2012. Print.

Glotfelty, Cheryll, and Fromm Harold, eds. *The Ecocriticism Reader: Landmarks in Literary Ecology*. Athens: U of Georgia P, 1996. Print.

Huggan, Graham, and Helen Tiffin. *Postcolonial Ecocriticism: Literature, Animals, Environment*. London and New York: Routledge, 2010. Print.

Lopez, Barry. *Arctic Dreams: Imagination and Desire in a Northern Landscape*. New York: Bantam, 1986. Print.

Midgley, Mary. *Animals and Why They Matter: A Journey Around the Species Barrier*. Harmondsworth: Penguin, 1983. Print.

Nash, Roderick. *The Rights of Nature: A History of Environmental Ethics*. Madison: U of Wisconsin P, 1989. Print.

Shepard, Paul. *The others: How Animals Made Us Human*. Washington, DC: Island Press, 1996. Print.

Tallmadge, John. "John Muir and the Poetics of Natural Conversion." *North Dakota Quarterly* 59.2 (Spring 1991): 62–79. Print.

Chapter Twelve

Ecocriticism beyond Animist Intimations in *Things Fall Apart*

Senayon Olaoluwa

A discussion of ecocriticism of the Global South stands to benefit from an acknowledgment of the environmental consciousness which courses through Chinua Achebe's *Things Fall Apart* (1958) with nuances that render it an embedded category in an otherwise overly anti-colonial text that served to decisively launch written modern African literature in English. The text transcends the limitations of western paradigmatic environmental thought and serves as an exemplum in the illustration of the complex environmental discourse of the Global South. This chapter explores the various layers of ecological concerns that find articulation in the text. It argues that an understanding of the ecological consciousness in *Things Fall Apart* must begin with a rejection of the reductionist view of cultural colonial frameworks that do not take into account the material, environmental, and spiritual aspects of African animistic rites and traditions. I attempt to affirm that the progressive moments in *Things Fall Apart* which endorse biodiversity conservation are those that are embedded in rites and tradition. Ultimately, I seek to explode the oppositional juxtapositioning of the discourses of modernity and tradition by ascribing progress to tradition, while associating modernity with ecological depletion through a critique of Christianity as portrayed in the text.

Theorizing ecocriticism in the African context is a complex cultural practice that has to take into account the diverse material and socio-cultural contexts in Africa and also undertake a critique of modernity that has shaped Africa in singular ways in the Western imagination. Speaking about the enframing of Africa as a place constituted by the absence of time, civilization or humanity, Byron Caminero-Santangelo and Garth Myers write, "Given the history of this representation as well as the continent's heterogeneity, it is tempting to dismiss any representation of Africa as a place as a fantasy, and a dangerous one at that" (8). An ecocritical discourse of Africa as a place-in-the-world has to be careful in acknowledging the role that global/colonial narratives have played in defining the local and at the same time in articulating the vital imbrication of the human with culture, soil, and the larger world of other-than-human nature in dwelling in place that is unique to the local. Caminero-Santangelo and Myers's succinct phrase "environment at the margins" captures the in-between nature of ecocritical discourses relating to Africa since these discourses occupy the "interstitial space of those inherited binary divides" (*Environment at the Margins* 15).

This dialectic between change and continuity at the heart of African environments is best represented by the art of storytelling. Scholarship in anthropology, science, and history regarding Africa has always privileged certain dominant forms of narratives thereby silencing native voices and worldviews. Such colonial narratives and the final truths they claim to present about African environments and social practices have been questioned by postcolonial ecocriticism. Classic works like Ngũgĩ wa Thiong'o's *Decolonizing the Mind* (1986) speak about the importance of language, both written and oral, in restoring multiplicity and complexity to the discourse of Africa. Ecocriticism emerging from the Global South also emphasizes the power of counter-narratives that employ the medium of debate and dialogue to offer more powerful stories about the African environment. These counter narratives affirm that awareness regarding the right attitude towards nature is essentially constitutive of African epistemic and cosmological practices. The view thus gives validation to the assertion that "to suggest that postcolonial ecocriticism is new is to give a normative status to ecocriticism's institutional origins without questioning the limitations of its foundational methodologies and focus" (Cilano and Deloughrey 73).

Obviously, *Things Fall Apart* is an informed engagement with the mementoes of western colonial adventurism, which was preceded by slavery on the African continent. However, a more productive reading should affirm that beyond the obvious critique, the text provides other informed reflections of the modern and postmodern epochs through the various encounters which argue against animist reductionism. Sophia Samatar asserts that beyond the centrality of anti-colonialism in the text is the propriety of admitting that there is a display of an "unflinching presentation of a pervasive and global reality: the condition of modernity" that is crucial to the understanding of the text (61). In a similar vein, Simon Gikandi's view as a libratory critical precursor to Samatar's affirms that in addition to being a paradigmatic critical trope of anti-colonialism: "Achebe's novel shifted the idea of Africa from romance and nostalgia, from European primitivism, and from a rhetoric of lack, to an affirmative culture" (8). The said "shift" mediated in the text is in the rejection of the uninformed thinking of the colonial District Officer whose ignorance leads him to describe the Umofia society at the end of the novel as the "*Primitive Tribes of the Lower Niger*" (Achebe 166).

In thinking about the word "primitive," we are compelled to ask what suggestion does it make of the anthropological description of African life in colonial times as at best animist and lacking any merit of reason? And how did such assumptions inform reductionist receptions which continue to interpret the text as concerned only about the past? A peep into the world of anthropology reveals it was a nineteenth-century intellectual invention intended to serve colonial interests. Led by Tylor, anthropologists categorized the colonized as the "exotic" in order to establish an alibi for their pacification as the "primitive" (Guthrie 106). In his reflections on why western colonial novels on Africa tended to misrepresent African characters as undeserving of sustained attention, Thomas Lynn says: "As for (European) novelists, focusing sustained psychological attention on individual Africans would spoil the mystery, and, in any case, one can scarcely escape the conclusion that these and other European writers considered Africans to be not interesting enough for focused interrogation" (53). What is more, the primitivism of the colonized tends to be articulated in the agency of binarism which often represents Western colonization as the transcendence of the colonized. One sure way in which this has played out in social theory and anthropological scholarship generally is the categorical affirmation of a rift between nature and society which has enabled technological advancement (Hornborg 21). While this view has held out for about two centuries as the Western world continues to seek justification for its actions against nature and the colonized, the implied isolation of nature has in turn produced a form of natural tyranny that today haunts the advanced world. In other words, the human estrangement from nature through a systemic process of untrammeled pursuit of technological advancement has produced a world order that is at best compromised and at worst threatened by an apocalyptic expiration in the face of mitigation efforts that are incommensurate to the havoc seen everywhere on the planet.

The emergent awareness of the devastating consequences of the blind pursuit of technological advancement without thought for nature or the human others, has resulted in an unprecedented revision of primitivism in anthropological scholarship in the twenty-first century. Therefore in penitent remediation of the castigation of nature and cultures we now encounter anthropological treatises that seek a reconsideration of the original conception of the primitive as propounded by Tylor. One such revisionism to absolve societies whose unbroken affinity with nature previously showed them as backward in western critical rendition is to be found in the scholarship of Nurit Bird-David (S67). By distancing herself from the otherwise condescending position of modernist thinking, Bird-David seeks to make sense of cultural practices which had for centuries been undermined and derogated as primitive and therefore worthy of minimal attention, if at all. The crucial place that "ecologies" and "environment" occupy in the discourse above needs to be emphasized, seeing that their invaluable utilitarian essence, which was once despised in modernist thinking, is now being ironically courted in the so-called "modernist" return to the appreciation of Nature. This new paradigm undertakes the reformation and resemantization of the so-called "failed epistemology" of animism. Referring to this new articulation as "relational epistemology," Bird-David defines this new paradigm and talks about how these epistemologies "dividuate" the environment instead of "dichotomizing" it thereby "turning attention to "we-ness" which absorbs differences, rather than to "otherness: which highlights differences and eclipses commonalities" (Bird-David S78).

In a similar revisionist mode of anthropology at the turn of the twenty-first century, Alf Hornborg admits that since livelihood in contemporary times is increasingly hinged on interaction with the ecosystems, "entities such as plants or even rocks may be approached as communicative subjects rather than the inert objects perceived by modernists" (22). Reviewing Bird-David's work on animism, Stewart Guthrie raises a crucial question which further illuminates our understanding about the return to an advocacy of reconciliation with Nature across disciplines and the institution of agency for the reconciliation of contemporary postmodernist culture and epistemology with the same. According to Guthrie, it is crucial to ask ourselves "whether we imagine we have social relations with nonhuman things and events because we animate them or . . . [we] animate them because we have relations with them" (106). Invariably the question underscores the point of animism which seeks animation of natural objects as environment and ecology—flora and fauna inclusive, acknowledging in the process the mutual dependence of one on the other. This finds an even stronger eloquence in the contemporary institution of agencies catering to and advocating the recognition of fundamental non-human rights (Frost 39).

All this, moreover, provides a strong basis for the contention that the revisionist fervour of the concept of animism by anthropologists at the turn of the twenty-first century did not merely coincide with the rise in the advocacy for environmental and ecological preservation and protection across all disciplines. Indeed, the conceptual revisionism by Western anthropologists must have been contemplated as a decisive move towards the reconciliation of nature and culture. The move underscores the necessary return to cultural practices previously derogated as "primitive" in order to utilize their epistemic templates for the invention of a new paradigm of a healthy relationship between humans and the non-human others upon whom the continual sustenance and survival of the former depends. Therefore, whether we "animate things" because we have relations with them or not is at best besides the point, because it is in the quotidian interaction of humans and the non-human others that our humanity is affirmed.

Ultimately, the sense of return to animism for the animation of contemporary ecological consciousness and ecocriticism generally consists precisely in the logic of what Alf Hornborg while citing Latour, refers to as "symmetric anthropology":

> an anthropology that does not merely represent an urban, 'modern' perspective on the 'pre-moderns' in the margins, but that is equally capable of subjecting modern life itself to cultural analysis. For animism raises more questions about ourselves than about the animists. To begin with, is it really true that we, modern 'Westerners', do not animate the objects around us? And, to the extent that we are indeed Cartesianists, whence does this objectifying stance derive? Is it a product of our social organization, our education, our personal biographies? What are the consequences of objectification, for ourselves, for social relations, and for the environment? Finally, why are we asking these questions today, and what relation do they have to actual social and ecological processes in our time? (22)

The reflection is mediated by a series of questions which are instructive for the unpacking of the ecocritical values of *Things Fall Apart* beyond the temporal paradigms of the West. This perspective transcends the mere rejection of anthropological critique of *Things Fall Apart* by opening up new and fresh vistas for revaluation of animist moments. It is my contention, moreover, that these are mainly moments of rites and tradition which accentuate the confluence of environmental justice and social justice issues in the ecocriticism of the Global South.

To begin with, the polarization of the world into Global South and North is not a new phenomenon as it is preceded by the dialectic of tradition and modernity. Yet *Things Fall Apart* clearly instantiates the best moments of tradition which call into question assumptions about modernity's teleology of progress. Against this backdrop, the reading of *Things Fall Apart* must avoid the divisive paradigm of the "metonymic" and the "symbolic," an idea that Ato Quayson argues is helpful in grappling with tropes of representation and contradiction in the novel (125). For, by blurring the lines between both levels of reading, we are able to fully come to terms with the import of Gikandi's earlier statement about the paradigm shift from "European primitivism" to "an affirmative culture" in *Things Fall Apart*. If there is a sense in which this speaks to the notion of "symmetric anthropology," it realistically does more than this by creating the springboard for a reading of the text in the timelessness of African ecocriticism which sets no temporal limits because environmental consciousness is in the African sense constitutive of its epistemology. The qualification for this progressive epistemic understanding therefore calls for the humility of the uninitiated outsider to learn from the wisdom in the cultural practices of the text regarding healthy human-eco relations.

While there are various portals of entry to the discussion of environmentalism in *Things Fall Apart*, the Week of Peace will make a particularly productive reading because of the multiple illustrations that it provides. By declaring a week as sacred and one in which there should be absolute peace, the Umofia community forbids noise or disharmony of any kind. Today there are various forms of research devoted to the reduction of noise in our environment in order to mitigate its impact on atmospheric pollution, and it is in this very sense that the epistemic wisdom of Umofia displays a sophisticated form of environmental awareness. Additionally the declaration of a week as that of peace is a point of policy and regulation; it also goes to show how such logic resonates with the contemporary national laws around the globe against noise pollution in the atmosphere (Vanheusden 276). Where such laws are put in place, not only are there punitive fines, but also, the government endeavors to sensitize citizens and potential offenders. It is on account of this that findings continue to accentuate the external negativities of noise of whatever kind to human health (Loupa D135; Occupational Health 28). In other words, just as human inventions and occupational tools of the present age are regulated in operational terms in order to check the hazards of noise pollution associated with them, be they vehicular or aeronautic, the Umofia society also regulates, as a matter of law, the utilization of tools and other associated properties during the Week of Peace. This it does as a measure to keep at the barest minimum any form of pollution which in a more familiar term in this context is designated as desecration.

The foregoing discussion then brings us to the question of our relationship to objects—animate and otherwise—in *Things Fall Apart*. Or put differently, what environmental sense do we make of the teleology of observing the Week of Peace in the wake of Okonkwo's desecration of the week by violently beating his third wife Ojiugo, thereby causing a disruptive noise that jolts the psychology and tranquillity of the community? As the Chief Priest of the Earth Goddess explains:

> You know as well as I do that our forefathers ordained that before we plant any crop in the earth we should observe a week in which a man does not say a harsh word to his neighbour. We live in peace to honour our great goddess of the earth without whose blessing our crops will not grow . . . . The evil you have done can ruin the whole clan. The earth goddess whom you have insulted may refuse to give us her increase, and we shall all perish. (Achebe 24)

The explanation of Ezeani, the Chief Priest, is replete with anxiety over good harvest and the scary destruction of the entire community. Okonkwo's commission of disruption to what I term the aesthetics of serenity and silence in the rites of the Week of Peace has various implications. Among these is the possible refusal of the earth goddess to sanction good harvest. This is in turn tied to the implication of nature's withdrawn largesse and how this translates into disaster for both humans and non-human others.[1] It is additionally so because "environmental crisis [as caused by Okonkwo's disruptive masculine violence] is not just a scientific and economic matter but involves cultural, ethical, and aesthetic decisions" (Matthewman 36). Besides the animist logic of affirming belief in the existence of a spirit behind the abundance of the earth, there is also a sense in which the week-long regulation of abstinence from noise and any form of violence proves an observance of honor to nature and cessation from the cultivational violence of farming, hunting, and other similar activities associated with people of the agrarian age represented in *Things Fall Apart*. The week-long serenity is thus analogous to some of the annual national vacations observed in contemporary times around the world like Christmas and New Year, after which people swing back to action in the workaday world feeling rejuvenated. Most significantly, the week-long observance of peace illustrates an African slant of ecocriticism which should among others be concerned about "what a society's assigning of significance to nature . . . reveals about both its present and past" (Vital 87).

The socio-cultural construction of climate change is a recurrent issue in *Things Fall Apart*. According to Luque et al., "climate change, [is] seen as a socially constructed anticipation of natural disasters and a future-risk that plays out in present politics" (738). The "politics" of disruption to the Week of Peace and the intervention of the Ezeani for that matter constitute the basis for the despair over Okonkwo's violent action which is capable of resulting in the unsettling experience of climate change. This is the import of the likelihood of the goddess of the earth refusing "to give us her increase, and we shall all perish." It is also because the "anticipation" is "socially constructed" in Umofia that Okonkow's actions are seen as capable of bringing woes to the entire clan. Any wonder then that in contemporary times such anticipation often results in "the emergence of new modes of governance in cities of the global south" (Luque et al. 738). This is why it is interesting to know that there is no rigidity to the law of the Week of Peace in the text, as it undergoes review from time to time. From place to place, it is said that people continue to review it, making it less stringent. In Obodoani, for instance, the law is reviewed so that people who die during the sacred week can be given a befitting burial instead of being thrown into the evil forest (25), which may result in the outbreak of pestilence. A similar instance of review of policy on the Week of Peace is the reversal of the initial capital punishment associated with it in Umofia (25), seeing that the prosecutorial process ended up breaking the peace.

At yet another level, the implications of the refusal of the goddess refusing to give increase consists in the manifestation of debilitating natural disasters which have the capacity to expose both humans and ecology to waste and destruction beyond redemption. The clan is already introduced to the enormity of the effect of climate change on food production and by implication human survival. The narrative alerts us to the "heavy rains" that drown the yams (5); the irregularity and extremism of rainfall and sunshine find full expression in the memory of Okonkwo (18–19). The experience illustrates a lived experience in Umofia which can be construed as "climate extremes," to borrow a phrase from Marvin et al. (444). The text offers another narrative context in which we are faced with a folktale which further underscores Umofia's epistemology as one that has implications for both human and non-human others. The nuances of this epistemology inspire awe for the pre-colonial world that it represents. A good example of this epistemic sophistication can be found in the story of the feud between Earth and Sky told to Nwoye by his mother (43).

The symbolic complexity of the tale speaks presciently to the inclusive agency that frames the conceptual construal of ecocriticism. This finds amplification in Timothy Clark's observation that: "Ecocriticism . . . does not write as if human beings were the sole occupants of the planet and must open itself to a space in which fundamental questions about the human place in nature are at issue" (5). This statement brings into focus the important role that natural space plays in human survival. The folktale clearly explains why in the crisis predicated on the quarrel of natural elements like Sky and Earth, humans are not only implicated, they are the worst hit. The climate extremes that manifest in form of a seven-year drought have devastating consequences not only for the living humans, but also for the dead since the rain-starved earth has turned stony and does not allow the burial of the dead. The epidemiological effects on the society are equally horrifying and better imagined than experienced. In the end, it still takes the intervention of an other-than-human agent, the Vulture to bail humans out. At one level, we can construe Vulture's role as a way of framing Nature as benevolent because of the Vulture's selfless plea that moves the Sky to compassionately grant the life-giving rainfall. At the other level, it is extremely important in this context to underscore the helplessness of humans in the face of natural disasters, since some occurrences which are not instigated by humans can nonetheless have far-reaching consequences for humanity. Of what significance then are the sacrifices and digressions in the Vulture's return journey to Earth? The obvious answer is that the Vulture was attracted to the sacrificial fire; but equally convincing is the argument that the Vulture deserves a worthy reward of the entrails of the sacrificial object, which is another way of illustrating the confluence of environmental and social justice. It is in this sense that the tale proves the logic of replenishment through which ecocriticism and other similar environmental intellections espouse an attitude of giving back. Beyond this, the folktale proves the assertion that stories that teach humanity and morality find reinforcement in "variant symbols in nature," demonstrating "that respecting nature is corollary to these factors" (Zolfaghrkhani and Shadpour 212).

If the gestalt of this can be perceived as response to the disruption of the cultural aesthetics of peace by Okonkwo, the anxiety expressed in its wake must be read as proving—again in a prescient way—the connection between climate change and food production. This much is demonstrated in the lousy harvest of the year in which Okonkwo attempts his first share-cropping, a situation that does not only provoke despair but also instigates suicide in another farmer (18–19). The extremes of weather portend nothing but disasters. In the words of Hans Marvin et al, there are "potential direct and indirect effects of such extremes as well as other severe weather and hydro-meteorological events on the occurrence of hazards in food produced by various agricultural systems" (444). In the prevention of lousy and unstable weather, the goddess of the earth symbolizes the agency of the non-human others. The human interaction forged between them and the epistemology of worship serve as the lens for visualizing how the formulation of policy and maintenance of atmospheric order against pollution privilege the agency of other-than-human elements. It is in this sense that interdisciplinary discourse continues to be disposed to thoughts around "the need to bridge the Cartesian divide between people as active subjects and inert passive objects, to better reflect how things provoke and resist human actions through their secondary agency" (Sillar 367).

Yet at other levels, environmentalism in *Things Fall Apart* is not exclusively about the Umofia held in thralldom to anxieties of natural disasters. This much is the case as we navigate the fond childhood memory of Unoka. Stable climate provides for the observance of the "carefree season" on an annual basis with children playing and singing at the sight of kites returning with the dry season. While it is a truism that play contributes immensely to the socio-psychological development of children (Ashiabi 199; Löfdahl 5), it is also productive to remark in the context of this discussion that *Things Fall Apart* reinforces the place of Nature and the environment in child development issues by showing how the whole process of play is enabled by a stable and benign weather. The benevolence of Nature and climate stability also find expression in the abundance that results beyond things for which humans labor. An illustration of this is the harvest of locusts which are said to visit Umofia at irregular interval of years (45).

In the next section of this chapter, I examine the disruption to ecology beyond remediation at the irruption of colonialism in *Things Fall Apart*. In the colonial world, since the idea of progress that drove the prosecution of western colonialism was synonymized with modernity, the formerly colonized spaces of the Global South in particular continue to interrogate the veracity of modernity's claim to progress. This interrogation finds summation in the now hackneyed phrase of the crisis of modernity. Or as Kerstin

Bowsher has put it, an assessment of the impact of modernity can at best be summed in the revelatory irony of "chimeras of progress [and] mirages of modernity" (97). Yet as theorists and critics of postcolonial theory and scholarship continue to debate the relevance of postcolonial studies in the twenty-first century, Robert Young asserts with affirmative certitude and profundity that the "postcolonial remains" (19). The import of this assertion consists in the explication that certain engagements of postcolonialism may have now turned obsolete and can appropriately be taken for dead. To that extent, the analytical tools and the social imaginaries to which they have been applied have now outlived their relevance and for that matter can as well be persuasively termed to constitute the dead "remains" of the conceptual category. The ambiguity value of Young's assertion is reinforced, nevertheless, in the additional meaning of "remains" which leaves us to ponder whether there are no other issues in contemporary times to which the analytical tools of postcolonialism can be applied. Of course, beyond the decomposing "remains" of postcolonialism, the postcolonial still "remains" with us in a twenty-first century that is overwhelmed by the anxiety of climate change and the consequent search for mitigation, an effort that finds privileged place in the age of the Anthropocene.[2]

Therefore, a productive approach to engaging with the pervasiveness of ecological destruction at the advent of colonialism in *Things Fall Apart* will be through an inclusive triangulation of major spaces in the text: Mbanta, Abame, and Umofia. The earliest mention of the coming of the white man in *Things Fall Apart* is in Mbanta where Okonkwo is observing a seven-year exile among his mother's relations. From literary studies to other fields of enquiry, contemporary scholarship is eloquent in its predisposition towards reinforcing the strong intersection of space and ecology as a way of fully accounting for the dynamics of ecological research (Lollar 45; Wylie 71). Against this backdrop, the Mbanta space is acknowledged as the space in which we encounter for the first time, a blow-by-blow account of the community's intimate interaction with the new religion of the white man. The discourse of space in Mbanta requires that we acknowledge the logic of biodiversity conservation, which though woven significantly around articulations of the sacred, leaves no one in doubt about the indigenous epistemology that places a high premium on ecological preservation and replenishment. It is a known fact across disciplines today that there is a strong link between forest preservation and biodiversity conservation (Yao and Roussel 63; Bush 471). The Mbanta community is well at home with this fact and makes it a constitutive element of its cultural practice. The preservation of the Evil Forest for the performance of various rites is an indication that here is a community that exhibits sophistication in its organization of space and ecology. For instance, the decision to bury people who have suffered communicable diseases within the space of the evil forest is for the community a preventive measure against the epidemiology of these diseases, given the fact that the forest is located far away from each given community (119).

The spatial and ecological equation articulated in the rites of the evil forest however changes with the advent of colonialism which deceptively employs Christian proselytizing to disrupt the indigenous social order. At this point, it is apposite to examine the dissection between the logic of sacralization of spaces in African religious rites and western Christianity as encountered in the text. While there is nothing inherently sacred about spaces and it is through human agency that spaces are so designated (Day 427), the agency of rites in Mbanta clearly shows that moments of rites also yield themselves to the enforcement of biodiversity conservation through the mapping of the evil forest as sacred. The advent of Christianity in Mbanta however er undermines this cultural logic of biodiversity conservation since with the concession of the evil forest to the Christian mission we begin to witness a gradual but sure process of deforestation.[3] Shortly after the evil forest is conceded by the Mbanta community, "The next morning the crazy men actually began to clear a part of the forest and to build their house" (120). Colonial accounts of "how Europe underdeveloped Africa," to put it in the words of Walter Rodney (1972), are replete with the centrality of depletion of African forests and timbers for export to Europe to the detriment of Africa. If this undeniable fact reverberates in the political economy of colonialism, *Things Fall Apart* asserts its perceptiveness by demonstrating how the foundation for the depletion of African forests was laid in the seemingly innocent clearing of forests to build churches.[4] The vulnerability of wildlife and other ecological elements is to be subsequently evinced in the overzealousness of converts, one of whom is Okoli who audaciously kills and eats the sacred python forbidden by tradition. Although the unpardonable sacrilege results in Okoli's mysterious death the following day (*Things Fall Apart* 129), it does not take long to be convinced that ecological destruction and biodiversity depletion are coeval with Christian proselytizing in the text. In other words, while the African construal of sacred spaces enhances biodiversity conservation through forestation, Christian sacralisation of space in Mbanta translates into deforestation and biodiversity depletion.

Abame, another major community in the narrative, comes next in the spatial discourse of ecological depletion in *Things Fall Apart*. According to Obierika, the people have responded to the intrusion of the white man who rides his iron horse into their community by consulting an oracle which forecasts that the coming of the white man will bring doom upon them (*Things Fall Apart* 111). The white man is killed in an act of proactiveness; but the reaction of the British colonial system does not merely show gross disequilibrium in its logic of retribution. The entire community is wiped out and "Abame is no more" (110). In place of one man a whole community is phased out, and there is a sense in which this returns us to the question of the

intersection of social justice and environmental justice. The killing of Ikemefuna may have its moral deficits especially with the involvement of Okonkwo against the advice of Ezeudu, but it surely instantiates an indigenous
retributive system that strikes an equilibrium between the weight of an of-
fence and retaliation. Otherwise, the initial murder of an Umofia daughter by
Mbaino could have resulted in a war in which many people would have been
killed. But indigenous wisdom prevails and only Ikemefuna is lost in place of
the woman killed. Similar options could have been explored by the colonial
administrators, but instead the system wipes out Abame.

More importantly, the interconnection of human and natural suffering is
demonstrated in the tragedy of Abame. As Obierika recollects, "Their clan is
now completely empty. Even the sacred fish in their mysterious lake have
fled and the lake has turned the colour of blood. A great evil has come upon
their land as the Oracle had warned" (112). These lines illustrate the indigenous epistemology that establishes the human connection to the natural
world, thereby showing Africa's holistic sense of ecological consciousness.
This is why in the death and displacement of Abame people, there is a
corresponding suffering of nature which manifests as the disappearance of
the sacred fish from the lake and the bloody mutation of the river. Further, of
particular significance is the dislocation instigated by the attack on Abame
people and the dislocation of the fish as well. As Paul Fryer and Ari Lehtinen
explain, it is vital to specify "the geographical scope of diasporic relations by
examining the historical transformation of social and ecological bonds in
particular moments of community displacement" (31). The surviving citizens
of the dastardly colonial ambush on the Abame people on Eke market day
flee to Umofia thereby initiating a new formation of diaspora as an indication
of their displacement from their original homeland. This resonates with the
mysterious disappearance of the fish, an indication that human relations to
nature and the question of animation find clarification in knowing the bond
between them, so much so that an injury to one is an injury to the other. The
consequences of the miscarriage of social justice by the emerging colonial
power has produced a situation of human death and displacement in the same
sense it has resulted in a miscarriage of ecological justice, resulting in the
depletion of biodiversity conservation.

The ecological destruction of Umofia as the third community in the narrative of spatial triangulation in *Things Fall Apart* indicates that, "apart from
the church, the white men had also brought a government. They had built a
court where the District Commissioner judged in ignorance" (*Things Fall
Apart* 139). Actions like these demonstrate how the forest which was once
regarded as a sacred space by the locals becomes gradually denuded of its
biodiverse plenitude. Enoch's killing of the sacred python, like Okoli in
Mbanta, provides the primary impetus. The sacred python is distinguished
from other such top predators like the giant python by its relatively small

size. It is generally held among many African cultures as sacred. Of all snake species that commonly live around human environment, it is only the royal python that is non-venomous. What is more, its calm, rarely aggressive behavior towards humans must have additionally been responsible for the level of unparalleled veneration that it enjoys throughout such cultures in Africa. Needless to say, the royal python is easily the most elegant of common snakes going by its radiant multi-colors, which makes it a site of beauty/ aesthetics. Possibly attracted to human homes because of the mice, its favorite prey, the snake's mild manner reputation inheres in its being able to live with humans in their homes for months, if not years without causing any harm in the process.[5] When all these known facts are put together, the logic of veneration is illuminated. For while other snakes are daily attacked and killed by human beings for their venom and the potential death they spell for humans, the veneration of the sacred python serves as a control and balance mechanism in the indigenous mode of biodiversity conservation. As Fakrul Alam has argued, "among the reasons he [Achebe] wrote *Things Fall Apart* was to show a continent where, especially before colonial invasion, people lived in a complex, diverse, and fruitful relationship with the African environment" (40). The resolve to deplete the sacred python in their habitat illustrates how the otherwise enviable harmony is impacted in a way that compounds human-eco relations at the advent of western colonialism and Christianity.

When all these progressive values are undermined and designated as pagan and animist practice, the consequences can be devastating as found in the fate of Okonkwo. In this sense, it is arguable that there is something cumulative in Okonkwo's resistance to the institution of Christianity and colonial government in Umofia. His resistance stems from multiple factors, an important one being Enoch's sacrilegious and intentional killing of the sacred python. Enoch's initial killing of the python that goes unchecked gives him the impetus to unmask the masquerade. It is the attempt to check the further excesses of the overzealous Christian convert, whose misdemeanor is bolstered by the extremist white missionary Mr. Smith's intolerant interpretation of the Bible, that results in the larger crisis in which Umofia is engulfed in the end. Put differently, the razing down of the church building and the subsequent incarceration of Umofia elders and the ultimate beheading of the court messenger by Okonkwo begin with the depletion of the Umofia forest and its biodiversity. This deserves emphasis because today the royal python comes in the long list of endangered animals. However, while it is commonly so in the West on account of its vulnerability to top terrestrial predators like the red fox (Bryant et al. 81), the increasingly alarming disappearance of the sacred python in most African communities today is as a result of indiscriminate Christian alimentary practices which declare all things edible. This attitude in a sense references Acts 10: 9–15 in which there is an express instruction to Peter in a vision to disregard any previous injunction to discriminate against animals including "creeping things," but to just "Rise . . . kill, and eat" (King James Version). That this has an implication for biodiversity in Africa is a cause for worry, as otherwise sacred animal taboos continue to be undermined daily. The place of Christian faith sometimes contradicts the principles of biodiversity conservation, a question that continues to preoccupy scholars of theology. As Lynn White argues in his famous essay, "The Historical Roots of our Ecological Crisis," the invention of a narrative of beginning in the Christian story of creation authorizes a binary between man and nature in a sense that affirms "it is God's will that man exploit nature for his proper ends" (52). Not only does such affirmation brutalize our orientation towards the non-human others, it also accounts for why humanity has become its own worst enemy, since in most cases, the brutality towards nature ultimately robs off on human existence. When taken to the extreme, as is often the case, the binary opposition invents otherness among humanity in order to affirm difference along racial, ethnic, religious, and ideological lines. The devastating consequences are inscribed both in ancient and recent histories of wars, conquests, genocides, and sectarian fundamentalism.

Ultimately, the informing Christian theology for the prosecution of Western colonial modernity in Africa, as elsewhere, was fraught with contradictions, which calls to question the whole idea of progress associated with it. As has been illustrated in this chapter, there is a sense in which the performance of colonialism refracts from its affirmations of progress because in its efforts to assert authority, colonialism violates the extant cleavage between nature and society in Africa. Yet, looking back at those preceding moments of tradition discussed in the text, we cannot but ascribe progress to tradition while contesting modernity's claims to progress. It is in this sense that tradition turns modernity on its head in a reversal trope. By instituting a new socio-cultural and political order that denigrates indigenous modes of human interaction with nature, the labelling of African cultural practices as animist and primitive fails in its persuasion to project a higher and more sophisticated realm of social and epistemic consciousness. It is in this very sense that *Things Fall Apart* exemplifies an African epistemic order which can be illuminated through the ecocritical examination of the Global South.

NOTES

1. This is central to my argument in "In Connivance with Nature: Inter-Faith Crisis and Ecological Depletion in Helon Habila's Measuring Time," *English Studies in Africa* 54.1 (May2011): 73-87.

2. See Will Steffen et al., 2007.

3. The seed of deforestation sown in colonial disguise of missionary church building continues to blossom unfortunately today in Nigeria with various stories about prophetic instructions from churches to members and sometimes communities to cut down the three that do not

only pave and adorn their compound, but also contribute to the sustenance of healthy oxygen production. Usually such instructions are revelations about how such trees have become coven for witches and wizard, thereby constituting agents of darkness. In their extremism, such prophesies can also take their toll on spaces of national importance. An example in point was the controversies that trailed the cutting down of trees that once adorned the Nigerian National Stadium in Surulere, Lagos because of an alleged prophesy that the trees were already compromised as meeting point for witches and wizards. See Ben Efe (2013).

4. Kofi Awoonor's poem "The Cathedral" illustrates the sense of communal loss that attends such act of ecological depletion in the name of building church structures.

5. This was an intimate experience of the author in his growing up years as a schoolboy in his village. A royal python had sneaked into his room, making itself comfortable under his wooden bed which was laid flat on the ground. The snake fed fat for months on mice in the room from its hiding place—which doubled as the author's bed—without the author knowing. Yet by the time the author sensed one night the rattle from under the bed was stronger than that of mice in the room, he checked only to discover that it was a royal python. Rather than being frightened, he was calm knowing, that it was not poisonous, although he had to find a way of removing it from the room.

WORKS CITED

Achebe, Chinua. *Things Fall Apart*. London: Heinemann, 1958. Print.

"Aircraft and Vehicle Noise Linked to Health Problems." *Occupational Health* 61.8 (2009): 28–28. Print.

Alam, Fakrul. "Reading Chinua Achebe's Things Fall Apart Ecocritically." *Bhatter College Journal of Multidisciplinary Studies* 1.1 (2010): 40–50. Web. 23 May 2014.

Ashiabi, Godwin. "Play in the Preschool Classroom: Its Socioemotional Significance and the Teacher's Role in Play." *Early Childhood Education Journal* 35.2 (2007): 199–207. Print.

Awoonor, Kofi. *The Promise of Hope*. Lincoln: U of Nebraska P, 2014. Print.

Bird-David, Nurit. "'Animism' Revisited: Personhood, Environment, and Relational Epistemology." *Current Anthropology* 40.Supplement (1999): S67–S79. Print.

Bowsher, Kerstin. "Chimeras of Progress, Mirages of Modernity: Abel Posse's El Inquietante Día De La Vida." *Modern Language Review* 100.1 (2005): 97–112. Print.

Bryant, G. L. et al. "Tree Hollows Are of Conservation Importance for a Near-Threatened Python Species." *Journal of Zoology* 286 (2012): 81–92. Print.

Bush, Terrence. "Biodiversity and Sectoral Responsibility in the Development of Swedish Forestry Policy, 1988–1993." *Scandinavian Journal of History* 35.4 (2010): 471–98. Print.

Caminero-Santangelo, Byron, and Garth Myers, eds. *Environment at the Margins: Literary And Environmental Studies in Africa*. Athens: Ohio UP, 2011. Print.

Cilano, Cara, and DeLoughrey, Elizabeth. "Against Authenticity: Global Knowleges and Postcolonial Ecocriticism." *Interdisciplinary Studies in Literature and Environment* 14.1 (2007): 71–87. Web. 12 June 2014.

Clark, Timothy. *The Cambridge Introduction to Literature and the Environment*. Cambridge, UK: Cambridge UP, 2011. Print.

Day, Katie. "The Construction of Sacred Space in the Urban Ecology." *Crosscurrents* 58.3 (2008): 426–40. Print.

Efe, Ben. "Witches did Not Make Me Cut Down Stadium Trees." *Vanguard* 29 September 2013. Web. 20 May 2014.

Frost, Laurie Adams. "Pets and Lovers: The Human-Companion Animal Bond in Contemporary Literary Prose." *Journal of Popular Culture* 25.1 (1991): 39–53. Print.

Fryer, Paul, and Ari Lehtinen. "Iz'vatas and the Diaspora Space of Humans and Non-humans in the Russian North." *Acta Borealia* 30.1 (2013): 21-38. Print.

Gikandi, Simon. "Chinua Achebe and the Invention of African Culture." *Research in African Literatures* 32.3 (2001): 3–8. Print.

Guthrie, Stewart. "On Animism." *Current Anthropology* 41.1 (2000): 106-07. Print.

Hornborg, Alf. "Animism, Fetishism, and Objectivism as Strategies for Knowing (or not Knowing) the World." *Ethnos* 71:1 (2006): 21–32. Print.

Löfdahl, Annica. "'The Funeral': A Study of Children's Shared Meaning-making and Its Developmental Significance." *Early Years: Journal of International Research and Development* 25. 1 (2005): 5–16. Print.

Lollar, Karen. "Binding Places and Time: Reflections on Fluency In Media Ecology." *A Review of General Semantics* 69.1 (2012): 45–54. Print.

Loupa, G. "Case Study." *Journal of Occupational & Environmental Hygiene* 10.10 (2013): D135–D146. Print.

Luque, Andreset et al. "The Local Government of Climate Change: New Tools to Respond to Old Limitations in Esmeraldas, Ecuador. " *Local Environment* 18.6 (2013): 738–51. Print.

Lynn, Thomas Jay. "An Adequate Revolution: Achebe Writing Africa Anew." *Critical Insights* (2011): 53–68. Print.

Marvin, Hans et al. "Proactive Systems for Early Warning of Potential Impacts of Natural Disasters on Food Safety: Climate-change Induced Extreme Events as Case in Point." *Food Control* 34.2 (2013): 444–56. Print.

Matthewman, Sasha. "Eco English: Teaching English as if the Planet Matters." *English Drama Media* (February 2012): 36–42. Print.

Olaoluwa, Senayon "In Connivance with Nature: Inter-Faith Crisis and Ecological Depletion in Helon Habila's *Measuring Time*." *English Studies in Africa* 54.1 (2011): 73–87. Print.

Quayson, Ato. "Realism, Criticism and the Disguises of Both: A Reading of Chinua Achebe's *Things Fall Apart* with an Evaluation of the Criticism Relating to It." *Research in African Literatures* 25. 4 (1994): 117–36. Print.

Rodney, Walter. *How Europe Underdeveloped Africa*. London: Bogle-l'ouverture, 1972. Print.

Samatar, Sophia. "Charting the Constellation: Past and Present in *Things Fall Apart*." *Research in African Literatures* 42.2 (2011): 60–71. Print.

Sillar, Bill. "The Social Agency of Things? Animism and Materiality in the Andes." *Cambridge Archaeological Journal* 19.3 (2009): 367–77. Print.

Steffen, Will et al. "The Anthropocene: Are Humans Now Overwhelming the Great Forces of Nature?" *Ambio* 36. 8 (2007): 614–21. Print.

The Bible. Brazil: The Bible Society of Nigeria, 1988. Print.

Vanheusden, Bernard."Country Reports."*European Environmental Law Review*14.11 (2005): 276–79. Print.

Vital, Anthony. "Toward an African Ecocriticism: Postcolonialism, Ecology and *Life & Times of Michael K.* " *Research in African Literatures* 39.1 (2008): 87–106. Print.

Wylie, Dan. "Playing God in Small Spaces?: The Ecology of the Suburban Garden in South Africa and the Poetry of Mariss Everitt." *Journal of Literary Studies* 27.4 (2011): 71-90. Print.

Yao, Adou, and Bernard Roussel. "Forest Management, Farmers' Practices and Biodiversity Conservation in the Monogaga Protected Coastal Forest in Southwest Côte D'ivoire." *Africa* 77.1 (2007): 63–85. Print.

Young, Robert. "Postcolonial Remains." *New Literary History* 43.1 (2012): 19–42. Print.

White, Lynn. "The Historical Roots of Our Ecologic Crisis." (1967) 1991: 48–52 Web.

Zolfagharkhani, Moslem and Reyhaneh Sadat Shadpour. "An Eco-critical Study of Chinua Achebe's *Things Fall Apart*." *Journal of Emerging Trends in Educational Research and Policy Studies* 4.2 (2013): 210–14. Web.

Chapter Thirteen

Ecocriticism, Globalized Cities, and African Narrative, with a Focus on K. Sello Duiker's *Thirteen Cents*

Anthony Vital

I begin by stating the obvious[1]: those of us who read and write in universities, who research and publish, live lives shaped by modern cities. Whether we live in large cities, in towns of various sizes—or live in landscape that is truly rural—we exist within transportation, communication, and resource distribution networks, reliant on technologies sustained by an array of urban institutions (which, depending on the region of the world, operate smoothly or with frustrating unpredictability). Moreover, we make sense of the world by drawing on cultures that bear the imprint of the thinking humans require if they are to negotiate and reproduce modern cities successfully. Just as importantly, the actuality of urban existence, its material processes, in addition to its cultures, have formed the subjectivity from which our thinking, moment by moment, emerges. Cities, in this sense, represent a normal and obvious fact of our lives—no matter how upsetting or disturbing we may find our particular urban places—which we negotiate as we earn a living, build personal relationships, and seek what gratifies us. The question motivating this chapter begins to address this complex actuality for us as disciplined readers of literature alert to global environmental concerns: how do we read in a way that acknowledges simultaneously our modern urban realities and the fact that failure to grasp human life in relation to natural worlds can lead to damaging consequences—for both humans and the life-forms they encounter? This question becomes especially urgent in an era in which we, as species, need both to adapt to—and, if possible, to mitigate—the effects of climate change. The question, so daunting to face, raises such a wide range of issues that this chapter can take only the smallest steps to address it; and because ecocriticism with an urban focus has yet to develop, I am as much concerned with method as with conclusions.

As this chapter illustrates, I envision an ecocriticism that joins in multi-disciplinary conversation about environmental futures, contributing literary studies' special insights into language and its crafting. Such multi-disciplinarity is crucial: modern worlds take their form from a wide range of discourses, both specialist and non-specialist, and the quest for a less destructive path into the future will benefit from as much communication among representatives of these discourses as possible. To this critical reflection, ecocriticism seems well placed to contribute thought about human behaviors that, inevitably, combine social interaction with some or other relation to nature—and, in this context, cities call attention to their central importance.[2] While I write in this chapter as a South African, it is as a species that we are affected by this urbanized reality: it is now commonplace that for the first time in human history more than half the world's population are and will continue to be city-dwellers. What we know as "globalization" rests on a network of cities wrapping the planet, and African cities, with the worlds they anchor, participate in this network in ways both material and cultural. So I turn first to consider an approach to reading literature from within this larger network—and because space in this volume is limited, I name only briefest pointers to what would need to enter such reading to account for what makes African worlds distinct. The focus on African narratives, I hope, makes up for this imbalance.

GLOBAL CONSIDERATIONS

In what follows, I make some observations, briefly, to frame such ecocriticism:

1. All cities currently exist in some or other relation to the current global socioeconomic order. As this order rests on pasts influenced first by modern colonial activity emanating from Europe, fueled by capitalist adventure, and then by modern empire, dependent on a thriving industrial capitalism, again centered in Europe but joined by the United States, and, now, by what Arif Dirlik terms a "global modernity," with power centers in both the North and the South, it matters that ecocriticism recalls a city's relation to these processes.[3] It is important, too, that ecocriticism recalls the histories of anti-colonial, anti-imperial, and then postcolonial thought, as it works to understand the many different "natures" that such pasts have helped to foster. If empire produced colonial conservation practices and its print culture fostered the various kinds of dominative attitude toward nature explored by John Miller in his *Empire and the Animal Body* (2012), it also produced, famously, the forest as resistance site in the Kenyan liberation struggle and, in colo-

nized places broadly, such cultural resistance that Elizabeth DeLoughrey and George Handley bring into focus in their introduction to *Postcolonial Ecologies*.[4]

2. While "global modernity" brings with it both invitation and pressure to modernize, cities will be places where the tensions associated with modernizing will be keenly felt—especially as human migration into cities across regional and national boundaries brings with it both the creative opportunities and the hazards of ethnic mingling and especially as access to social power and position is seldom (if ever) granted equally across any social divide. Therefore discourses about "nature," which different cultural memories under stress produce, will be equally varied and will provide an important context for reading particular texts.

3. This urban life is, of course, studied within modernity's diverse academic disciplines, which, together, form one key element in global modernity's reproduction. In both the social and natural sciences, scholars explore the consequences of urban life and its growth. What matters here is that, fed by these discourses, ecocriticism needs to rest on the understanding that cities have never existed "outside" of nature, that city-dwellers draw into cities both nutrients and energy as well as material for their built environments, sometimes from near at hand, but now, in modern worlds, from extended trading networks of sometimes planetary scale.[5] The built environment, in turn, exists in place, a place that has location within a matrix of natural flows—and, so located, a city's built environment will play its part, along with any material effluent from city-dwellers, in altering these flows (in other words, "impacting on the natural environment"). Moreover, in modern worlds and urban environments especially, what we experience as distinctively local is in actuality an interplay of the local and global, with the dominant global civilization affecting life chances for all life-forms, including the human. Current urban subjectivity, especially among the comfortable, is formed by these material processes, which remain mostly accepted with, perhaps, only fragmented understanding. How to integrate the drama of human need and desire into a rich awareness of natural processes is difficult enough in modern worlds; the complexity of urban material reality adds massively to the difficulty. Yet developing such awareness matters and, in the context of literary interpretation, ecocriticism could serve this project. Such (re)interpretation would, in the most general view, amount to rethinking the social as the ecosocial—a sphere in which the processes of urbanism, natural and social, in all their layered complexity, take central place in the struggle to understand and improve human life.

4. I have suggested elsewhere ("Toward An African Ecocriticism" 87–90) that ecocriticism on a postcolonial planet might work best by focusing its analysis on discourse, with discourse understood in material terms and with literary discourse as distinct even as it intersects with other discourses repro-

ducing modern realities (both academic "specialist" and ordinary "non-specialist" discourses). Ecocriticism, while it focuses on language, would stay aware of the social relations affected by modernity's pasts, social relations that mediate relations with what we call "nature." With this historical awareness would come, inevitably, an awareness of the colonizing activities that have fueled modernity's development, and the role of cities in these activities, past and present. If ecocriticism pays attention to how language from the past bears traces of this development, it will not forget how its own discursive activity intersects with the present. Globally, academic culture—together with its discourses—would not be possible without significant capital investment, a capital investment that links academic culture to the global order's networks of production and consumption—and their ecosocial consequences.[6]

5. If the social and natural sciences can give insight into the processes involved in modern urban civilization's material existence on this planet that humans have been so successful in populating, it is the distinctiveness of literary discourse, with its shaping conventions and its attention to tropes and their interplay, that ecocriticism would highlight. All literature's crafting of language works on multiple levels to involve, in its naming of the world, the stress-filled dynamism of subjective life, with its expression in moral and sometimes political choice. What needs to be emphasized is that, in the modern world, this subjectivity expressing itself in literature is produced in large part by urban realities—material, social, and cultural.[7] An ecocriticism of the sort I envisage here would explore how, via the text, "city" and "nature" enter into subjective life and then would sharpen its critical focus by drawing on other specialist languages, those of social and natural scientists. Thereby readers would gain significant critical distance from inherited forms of meaning-making that, however plausible they might appear, may not be adequate to the tasks ahead. In the near and far future the impact of urbanized human life on our species and on the planet will need to absorb attention. Ecocriticism might thus assist in cultivating a version of the "scrupulous subjectivity" Edward Said writes of in "Reflections on Exile" (1984), one that makes more conscious our own participation in the pleasures and costs of modern cities, with an understanding of the past and alertness to an accompanying sense of responsibility to the future (184).

6. The strategy I suggest in this chapter involves acknowledging that every literary work, in its singularity, distorts and omits as well as reveals—while, at the same time, in its drawing from available cultural attitudes and literary conventions, it ties itself to a moment in social and cultural history. Reading would grasp richly what the text explores—and then would explore what the literature does not give us. Using the notion of inevitable limits, inevitable partial perspective, it would reach toward what the discourses of the natural and social sciences can reveal about "cities-in-nature"—bearing

in mind that both discourses and cities are always changing. Literary works have their own concerns—which, in the case of narratives, tend to be social—and the stronger the work, the more powerful and plausible those concerns will appear. Yet, it is precisely in their power and plausibility that they will work to draw attention from the materiality of ecosocial processes. Not expecting to develop solutions, such ecocriticism can nonetheless enter into multi-disciplinary conversation by illuminating literature's exploration of desire and its frustration, of felt need and its thwarting, as these are experienced in specific social worlds, exploration that may or may not find any resolution. Illuminating this subjective dimension can add significantly to what the objective sciences supply.

For an ecocriticism within and for Africa, the task would be to think from all that makes African worlds distinct, both in relation to worlds on other continents and within Africa—while also acknowledging points at which Africa participates in global realities.[8] Such ecocriticism would incorporate available natural and social scientific scholarship reflecting local concerns, while advancing understanding of the formal features of African literary discourse. It would in this way bring literary texts into conversation with understandings of African cities' ecological placement, their regional social and cultural differences, their multi-ethnic characteristics, their relation to regional rural worlds, their past relation to the specific cruelties and difficulties occurring during an insertion into modernity via European colonization, and their present uneven relation to the globalized economy. In relation to this economy, African societies have been creating their own African modernities and, in the cultural sphere, resolving, subjectively, attendant tensions (over, for example, social inequalities, the struggle for good governance, and the reconciling of past values and present, rural values and urban). An African ecocriticism would bring into play awareness of such material, too, as it sets in creative tension African literary texts and African ecosocial realities.

READING NARRATIVE

In turning to read literary text, I focus on K. Sello Duiker's first novel, *Thirteen Cents* (2000), while glancing for comparison purposes at Zakes Mda's *The Heart of Redness* (2000), Ngũgĩ wa Thiong'o's *Devil on the Cross* (Caitaani Mutharaba-ini 1980), and Doreen Baingana's collection of interlinked short stories, *Tropical Fish* (2005). Each narrative (or, in this last case, a sequence of narratives) concludes with a sense of limited personal or local possibility in the face of an uncertain or troubled social future. With important difference in *Devil on the Cross* (which focuses on the body, or, the "body-as-nature"), it is "nature" outside the realm of the human, not the narratives' cities and their social workings, that offer the characters that (albeit limited) sense of possibility. Yet all the narratives in their writing of nature, in their writing of cities, preclude awareness of the complexity of the various kinds of relation that exist between nature and city.

Thirteen Cents stands in fascinating contrast with *The Heart of Redness*, published the same year. In Mda's novel, the focalizer is the university-educated Camagu, ready to take his place in the urban heart of a new South Africa. Quickly disillusioned by the power dynamics that appear to be driving this urban world, he transfers his intellectual engagement to rural life, one that is still shaped by indigenous cultural understandings. The novel, by its conclusion, suggests the value of reconnecting with such rural, ancestral worlds, defending them in a flexible, postcolonial spirit of negotiation between cultural inheritance and modernity, as a way to confront a South African future made unpleasant and unpredictable by the workings of big capital.[9] Duiker's novel, too, articulates a disillusioned response to post-transition South Africa and, as with *The Heart of Redness*, it represents the urban world as an unappealing, confusing place, where money is power and power is bent on excluding those in need. What Duiker's novel adds, which the voice of Azure, his first-person narrator, delivers, is a sense of the city as experienced by a twelve-year-old "street child": a place of constant predation, sometimes violent, sometimes coating its desire to control and consume with a veneer of concern.[10] The narrative explores the resulting feelings of uncertainty, mistrust, and physical vulnerability in the boy and how he responds to these; yet it is his struggle for self-confidence that he will need for survival that drives the narrative to its conclusion. The child's voice communicates, as well, a plausibly limited understanding of the world. This, when combined with his concluding achievement of self-confidence, suggests, as I elaborate below, some of the novel's difficult ethical and political implications, deriving, it would appear, from the novel's sense of its historical moment.

Causes of Azure's lack of self-confidence are many and obvious. He suffers physically and emotionally from those he interacts with and, perhaps more damagingly, he is made aware continually of his status as a "boy" in a world of adults—even as he senses that this is a moment of personal transition—"almost a man, nearly thirteen years old" (1). Exacerbating this sense of inferiority is his increasing awareness of himself as quite alone, as needing to rely for survival on a sense of his own power. While he does enlarge—to some small extent—his understanding of his street-world, it is more important that he find resources within himself to boost his sense of being able to stand independently. He knows that the ambiguity of his appearance (of African descent but with blue eyes—and so not "black," or "white," or "colored") causes him difficulty on the Cape Town streets, scarred by the legacies of racism. He knows too that, as a street-child in a city dominated by money and the property it buys; he owns almost nothing material, needing to hustle to survive, selling his body to men. Yet, he also knows that he needs more than material subsistence. He can negotiate with some of the city's untrustworthy adults: from the adults who want sex, he insists they give as well as take. But his dealings with the gangsters that rule the streets teach him a sharp lesson. Both gangsters and pedophiles may take physically from Azure, yet the gangsters also exact "respect," reminding him violently of his actual powerlessness to bargain, even for subsistence. Owning little, he owns nothing unless he owns himself, which—and the novel in its ending under-scores this—he interprets as carrying within himself a sense of his own capacity for violence. As the narrative progresses, his reflective abilities increase, as does his rage ("destroy, destroy" and "I am getting stronger" appear as refrains).

Duiker noted the biographical origins of the novel in his friendship with street-children ("The Last Word"). Yet the novel is more than an adult's attempt to relay a child's experiences though a "coming-of-age" narrative.[11] In building the child's narration, Duiker draws on attitudes circulating in his historical moment. What is an impressive imagining of a child's mind also becomes a vehicle for exploring an adult response to post-1994 South Africa, with Azure's moment of personal transition recalling the moment of histori-cal transition. This latter should be empowering; instead, as this narrative "of a child" suggests, the moment is one in which, for Duiker, it makes sense to feel as an orphaned and homeless child in the face of history. Rather than seeking (as Mda's novel does) to find in rural worlds a compensation for disillusionment with the contemporary urban order, the novel stays resolutely engaged with what it senses as a predatory civilization ruled by untrust-worthy "adults"—gangsters, bankers, false friends and helpers—and shot through with inherited racial animosities. Yet references to Cape Town's natural world do carry a kind of saving significance. From the beginning, the narrative reveals the boy to be aware of a reality beyond the Cape Town in which he seeks to live. Even while naming some of the daily aggression and predation that Azure needs to negotiate, the first chapter sets out neatly what the narrative will develop as a contrast between a nature urbanized (its pi-geons, its rats, its sandwich bread) and a nature intuited as free (the ocean). It is the ocean that gives Azure a sense of healing and cleansing power as he bathes in it, especially after sex-work (30).

While the ocean appears mildly beneficial in the narrative's early sec-tions, as Azure experiences more and more oppressive violence, including a gang-rape, it becomes associated with violence as well as freedom, with, in other words, a liberating violence. A gang-leader may forbid him to bathe in it (73), but the ocean has already come to him (50–55). In an extended episode, while this gang-leader holds him captive—and during which he suffers the gang-rape—he has his first significant encounter with seagulls.

Locked on the roof, Azure watches with excitement as the gulls, suffering no such restriction, drive off the pigeons he scorns and fears for being informants to gang-leaders, enacting a tough masculinity not subject to predation ("seagulls have pride" [51]). Then, in front of one of these representatives of ocean, a large "man seagull" (54), Azure performs his own display of dominance, urinating on the gull's excrement, playing out behaviors crucial, he feels, for his survival. This small drama enacts attitudes echoed in his refrain, "I'm getting stronger" (55), and these attitudes shape the concluding apocalyptic vision in which the ocean rises up to destroy the city. For it is the whole city, including all adults and not only the gangsters, that inspires fear and rage—feelings reinforced by the words spoken before his return visit to the mountain by a "friendly" gang member: "Evil hides itself . . . . In the church, in banks, in town. That's why we have to destroy Cape Town. . . . That's how you fight evil. With evil" (138).

Accompanying this growing self-confidence is an increasing tendency toward introspection and, of these moments, his reflections on fear appear most significant. On his first visit to Table Mountain he expresses his distance from the "white people" he encounters, people who move across its surface as if they were its "owners": "they don't look at the ground. They only look ahead of them. That's why animals are always running away from them. . . . White people don't know fear and animals know that" (124–25). Yet, if, to his mind, a culture of owning—and, especially, owning "nature"— exempts white people from fear, those on the social margins can move with confidence by "learn[ing] to live with fear," as happens in the gang world's violent control of the streets. Azure appears to reject both paths as he reflects: "What does that mean [learning to live with fear]? It means grown-ups are evil . . . they use anything they can use and when they get it they still want more" (143). By the end of the novel he finds his own way of dealing with fear. Reflecting on the harshness of the world he is born into, a harshness he thinks of as matching the sun's, he comes to terms with fear through his own form of "ownership," one based on the visionary. For Azure, the vision is of nature rising up violently against the violent, hurtful city, destroying what is destructive (162).

This visionary sense of nature can be noted from the narrative's beginning. Azure's response to the seagulls echoes in a different register his response to the pigeons. In both cases, he interprets animals as spirit-beings, as existing on the same existential plane as he does, the pigeons as sinister representatives of spirit subjugated by, and therefore in the service of, urban power, the seagulls as representatives of ocean, a spirit of healing that is mild but also potentially destructive. It is on the mountain that he experiences spiritualized nature in terms, that points back to ways of living prior to the appearance of a colonial modernity on the African continent (though, as Shaun Viljoen notes, the vision assembles biographical and popular culture elements as well [xvi]). Unlike the ocean, which has value as what lies entirely beyond the human, the mountain has value for how it allows such vision by being a space for humans that exists nonetheless outside the city. Here Azure, through trance and the medium of fire, encounters the spirits of the Cape's earlier inhabitants, first its animals and then a woman (she calls herself Saartjie, suggesting Sarah Baartman and recalling the historical depth of the cruelty in this boy's world). The final apocalyptic vision of nature's power, a vision that affords him a sense of connection with both old African ways of knowing and nature's power, serves as a culminating moment. It offers Azure psychological liberation from fear and a sense of how, in this world as harsh as the sun, he owns himself, at least.

Yet, what is one to make of this ending? There is no consciousness outside the boy's that can be turned to for guidance: the moment's meaning for him is its meaning, a meaning the adult writing presents as valid, even if he may not endorse it. The vision's solution to the boy's life situation, however disturbing for the homicidal rage it can be read as representing, makes sense within the terms set up by the narrative: the boy's marginal economic position (he cannot begin to articulate an industrial worker's perspective), his limited understanding of social history, his sense of aloneness, his sense that in this bewildering world it is only a self-reliance enacted as destructive violence that can help him survive. In the world left him by 1994, the friendships and alliances he finds are too unstable to be relied on and the uncertainty about the future that ends *The Heart of Redness* is also evident in this narrative. The social world in South Africa, post-1994, offers no scripts for systemic transformation other than the official version in line with the needs of capital, both local and global, supporting an obviously problem-filled status quo. Duiker has nothing to draw on to offer his character other than a liberation that is internal, reinforcing of his solitude and echoing the violence of the pasts that has produced his city.

Azure's attitudes reproduce, in extreme, the dominant neoliberalism, one maintained by those very gangster-bankers he seeks escape from. The historical moment of 1994 appears emptied of meaning and with it that history of social struggle embodied in modern revolution (traceable back to seventeenth-century Britain and its "Glorious Revolution" of 1688), a history that supplied the struggle against apartheid, as in much anti-colonial struggle, its language of democracy and rights. In this post-1994 novel, the world of "adults" who are "full of shit" (106) will never be displaced; it can only be entered in its own destructive spirit (though not joined). Existing struggles of communities, of workers, have no place. It is chilling to read these final pages (published in 2000) with Al Qaeda's 2001 destruction of New York's Twin Towers in mind. Thinking forward to such North American texts as Jared Diamond's *Collapse: How Societies Choose to Fail or Succeed* (2005) or the film *A Crude Awakening—The Oil Crash* (2007) suggests that Duiker is shaping his narrative in accord with a postmodern temporality haunted by a post-Cold War version of apocalypse, in which day by day the global economy lumbers on, driven by aggressive self-interest, falling apart here, reconstituting there (thanks to this or that exciting, promising technological advance)—yet haunted by images of its own violent end. In such haunting, either people hostile to this global order will destroy it (and the increasingly intrusive electronic surveillance by world powers, justified by the need to guard against terrorists, speaks to this haunting) or nature will, as socioeconomic imperatives modify the planet's ecosystems faster than they can recover.[12]

Duiker's narrative, of course, in its social focus, gives sympathetic voice to one who seeks remedy in visions of apocalyptic violence and does not concern itself with ecosystems—or Cape Town's relation to them. Its focus on a child, and one most dispossessed in this city,[13] produces a discourse far from the discourses of a city's professional and managerial classes. Moreover, its social focus displaces powerfully any thinking about Cape Town's life in its natural situation. While it depicts astutely the class and race dynamics that are Cape Town's inheritance from its colonial pasts, its depiction of nature is driven by the boy's understandable need for a meaningful space that is "outside" what Cape Town confronts him with. A moment on the mountain offers an emblem of this displacement. If the mountain offers Azure refuge and the possibilities of self-discovery (intruded on annoyingly, at times, by others seeking pleasure there), it also reminds him of the structures of adult authority that he climbs the mountain to escape. He is chased away from the reservoir he swims in and the moment registers as simply one more imposition (125). The complex history of water provision in Cape Town is eluded to (the fact that it was once deemed necessary to build local dams to supply Cape Town with its water) and then bypassed. Questions of conservation in this region, with its long dry summers, of the need to find additional water sources in mountains distant from the city to ensure a steady supply of potable tap-water, of expanding the use of "grey" water, of protecting the city's waterways and wetlands from a pollution that has multiple kinds of origin—questions of the kind faced by city planners and crucial for understanding this dimension of the city's metabolic life—from these our attention is drawn away ("Welcome to the Water"). So too does the narrative draw attention from other dimensions of this metabolic life, such as food and fuel supply or sanitation and other waste disposal requirements. Displaced, too, is the relation of Cape Town's infrastructure to the fragile ecosystem within which it is built and which, inevitably, as the city expands, gets increasingly disrupted.

A fuller reading of this novel would incorporate more of the social and natural reality with which Azure's narrative intersects and thereby feed the readers' imagination with a sense of what is at stake in this city's life.

Obviously, not all urban social problems have an ecological dimension; homelessness among children and sexual predation on children are examples of social problems needing urgent attention. While cities, necessarily, to be effective, need to compartmentalize ("departmentalize") their work, literary study, through its critical explorations, can encourage a seeing "with both eyes." Holding the social and ecological in mind simultaneously could contribute to promoting a culture that grasps reality as ecosocial, a culture that would then affect how specific problems are understood and addressed. Yet, reading the novel suggests, as well, some of what lies outside the scope of such "objective" realities. Azure has a need for a nature "outside" the city, for a nature that connects him with a pre-colonial culture, a culture that supplies a relation to the natural world very different from that fostered by the daily reproductive imperatives of global capitalism. Understanding need in relation to all else in the narrative draws attention both to the novel's subjective dimension and to the cultural and social dimensions shaping subjectivity. Such attention can add crucial material to objective accounts of ecosocial realities, enriching conversation in professional meetings and classrooms about our life in cities and towns and in the nature that sustains us.[14]

Reference to the cultural and social dimensions shaping subjectivity recalls the point that *Thirteen Cents* is not simply "about a homeless child." As noted above, Duiker, with Azure as focalizer, is elaborating responses to his social world. That he turns to nature in his narrative reproduces a move Mda makes in *The Heart of Redness*. Duiker's novel, though, engages head-on with the aggressively acquisitive individualism to be found in modern and postcolonial city streets (masked, where possible, by orderliness, temporary alliances, and—as the episode with a wealthy banker suggests—by politeness and a veneer of "high culture"). Whatever can be done for the ecosocial life of cities would need to recognize and counter the effects of this individualism (and its "gang member" variants), an individualism that is rooted in modernity but exacerbated, as Duiker's historical vision suggests, by the neoliberal tendencies that South Africa was liberated into in 1994. Under such social conditions, it is reasonable to expect that urban life and its struggles will produce a distorted understanding of nature in its relation to urbanism, one that blocks, rather than enlarges, an awareness of the city-in-nature. If Duiker's novel enacts this understandable distortion, so too does *The Heart of Redness*. As noted above, the narrative moves away from Johannesburg to focus its intelligence on the rural Eastern Cape, there exploring various issues, including both the environmental health of the place as it confronts the forces of large-scale development and the value of reconnecting with an African cultural tradition. This novel, too, finds value in a nature outside city-limits, one infused with cultural memory drawn from a time before colonial modernity. It, too, appears driven to do so in rejection of an urban life tied to globalism; and in turning to locate value in rural community, encourages a blindness to the present and past and the complex ways in which urbanism both affects ecosystems and generates inadequate images of the nature-human relation (including those of rural life as somehow "closer to nature.")

In developing this way of seeing the ecosocial life of cities, it is important to remember that the "ecological" includes us who read and write (a remembering that the abstraction of language can distract us from and even impede). Humans do not exist independently of nature's flows. These flows constitute our bodily existence, whether we live as hunter-gatherers or as residents of Johannesburg, and so it is crucial to augment abstract thought about urban metabolic life. Without human bodies, in all their complexity, there are no cities; without the natural systems to sustain these urban bodies, cities would be emptied of people.[15] As humans dependent now on globalized cities, we need to struggle to understand and come to terms with the intricate ways our bodies, and so also our minds, are involved in this urban life, constituted by constantly shifting global interconnections. As mentioned earlier, the narratives we read, even those that draw attention to the body, are likely to draw us away from attempting such ecosocial understanding, motivated as they will be by awareness of largely social and psychological problems and the need for their resolution. Nonetheless, such narratives will speak to subjective issues crucial to an ecosocial understanding that includes our involvement in cultural dynamics. To illustrate, I end this section with a brief reference to Ngũgĩ's *Devil on the Cross* and Baingana's *Tropical Fish*, which, in addition to focusing on the body's life in urban worlds, suggest the possibility of an ecosocial order—and life for the body—very different from the current reality.

Ngũgĩ foregrounds the body—and specifically, a woman's body—in a novel that exposes contradictions—and outright lies—in Kenya's "independence." Self-conscious in its modernist artifice, yet rooted in the *gĩcaandĩ* genre in Gikuyu oral culture, the narrative's fragmented representation of modern Kenya emerges from its narrator, a self-proclaimed Prophet of Justice, with characters (even Warĩĩnga, the protagonist and most fully developed) existing as types in his vision of the national drama.[16] This narrative, as with Duiker's, concludes in a moment of violence, with Warĩĩnga fugitive from the forces of the state as they seek their corrupt brand of justice, one that rests on unrestrained exploitation and suppression of dissent.[17] If heroic Warĩĩnga takes her stand against these forces shaping modern Kenya, Ngũgĩ's narrator, his voice informed by the traditional ethical and political values that inspired the liberation struggle against British colonials, shares in that heroism for telling Warĩĩnga's story, supplying Kenya's people critical purchase on the inner workings of their daily reality. Within this frame (broadly defined) Ngũgĩ inscribes "the city," with urban spaces touched on lightly to illustrate points revealed in the national drama. So, the peasant woman Wangari's story of alienation from the capital city of her independent homeland is pointed by how she describes Nairobi; Warĩĩnga's opening helplessness is underscored by her description of the confusion and hostile atmosphere of Nairobi's streets, by the sexual predation in Nairobi's office spaces. At a key point in the narrative (once she has had her eyes opened) she depicts as horrific the inequalities evident between those who live on Ilmorog's Golden Heights and those who live in the slums (130). Though the narrative has no interest in developing an urban social history (in the realist sense), the city emerges as social product, one that implies the shaping power of colonial—and now, neo-colonial—interests.

"Nature" enters the narrative in the same way, as filtered through the awareness of the characters. Mostly, as is fitting in a work against neo-colonialism, the narrative figures nature as resource: whether, for example, land and animals for Wangari, the peasant, or material for enrichment for the capitalist "thieves and robbers." When it presents the natural world outside the realm of economy, it does so in the context of the seductions of bourgeois romance that Warĩĩnga experiences—after her co-op is lost to capitalist land development (225). This short-lived co-op, and Warĩĩnga's schemes for solar cookers, suggest ways in which modern cities and nature might exist in harmonious relation, one in which nature plays the beneficent role it does in traditional tales, as an economic resource that is communal, sustaining the welfare of all society. Such revision of the social order, however, remains only a suggestion—and, with organized opposition to the status quo represented as broken or contained, it is only in personal life that the ideal of a self-confident African self-reliance survives. In Warĩĩnga's sense of her body, her rejection of Western standards of beauty and "return" to a traditional body, strong in work and in self-defense, and carried with pride, the narrator offers the seed of a better social future. The economic conditions enabling such a future may be out of her control, but her body, freed from Western cultural influence, remains hers. This body she defends, in her act of violence, from "takeover" by the same member of the capitalist class who had lured her, when young, into leaving her education—and the potential for independence—to become his mistress (before he cast her aside when pregnant).

While the adult Warĩĩnga's liberated but fugitive body, defended by strong will, historical understanding, and force when needed, may be all the "nature" that can be protected from predation, the co-op that fosters her transformation remains in the Prophet's telling the only positive locus of social resistance in a world dominated by transnational capitalism. That the co-op disappears so swiftly from the narrative signals how, in this narrative, the interests of an all-powerful capitalist and neo-colonial order define nature's relation to urban worlds. The class benefitting from this order, the local "thieves and robbers," who are rewarded handsomely for doing the bidding of their metropolitan masters, has and would have no interest whatsoever in reviving an African communalist relation to the natural world. The cities that might grow from such a change, with the moral and economic forms shaping that growth, are, in this narrative, unimaginable. So Wariinga's survival, the survival of her independent spirit, and the vision of liberation that she embodies in the Prophet of Justice's telling, remain the only potential for seeding a future in which an African modernity aligns with its cultural roots. Yet, in this schematic and, in that sense, idealist narrative, the novel is driven by its political motives to overlook how urban mediation complicates social and individual relation to nature. In the co-op, the novel may offer an ideal of material distribution, yet the issues surrounding the supply of materials needed to provision and sustain an industrialized city remain outside its focus. While it could be argued that the co-op idea might well inform the provisioning of a city with food and water, for example, and even the manufacturing of industrial goods, the same would be hard to argue for the extractive industries needed to produce an industrial built environment. Historically, these industries have been implicated in deeply troubling ecosocial realities and will not be easily transformed—especially on the global scale that the narrative acknowledges is powerfully determinant. Global interests will ensure that "their" metals and oil keep flowing; social and ecological costs will remain externalized onto those least able to resist.

Ngũgĩ's male narrator locates in women's lives and women's bodies a site of resistance to the subjection that makes a mockery of Kenyan "independence," a subjection to an alien culture promoting unrestrained greed. The narrator may view women's lives with sympathy, aware of the shift in their economic position from producers central in village life to employees in a cash economy directed by men, but he views it from the outside. Baingana's short stories, in realist mode and without Ngũgĩ's alienated confrontational spirit, reflect a woman's apprehension of women's lives and bodies. Featuring moments in the lives of three sisters coming into adulthood, the stories reflect difficulties in the experience of middle-class women in Uganda—and yet, while the difficulties focused on are social and psychological, there is a moment at the collection's end, when the narrative suggests a very different sort of experience in an as-yet unrealized ecosocial order. Aside from this concluding moment, the narratives depict struggles for an adult self in a post-independence social world—once again depicted as dominated by men (local and Western). In a variety of settings—family, school, university, and work place—the stories track their characters' attempts to integrate emotional and sexual satisfaction into that sense of self. References to a natural environment in these stories with an urban setting tend to be of slight significance; yet, in the final moment of the last story, the writer turns to nature for a resolution. The central character notes, with wistful hope, how people are transforming the urban landscape, replacing colonial-style lawns (reproduced fervently in the United States of America, from which she has recently returned) with productive urban gardens, a form of interacting with natural processes that accords with African cultural memory (146–47). In this postcolonial moment, the future she faces appears less fraught with the difficulties from her past; she faces it with the resolve of the newly returned, to "dig deep down into this mud with her bare hands" (147). Placed as it is, at the narrative sequence's end, the momentary observing of the gardens cannot erase a difficult past or remedy all that she will confront as she negotiates a return to her urban home; yet it does offer her both hope and a metaphor for engaging with future difficulties, personal and social.

One can think of this moment as one more instance of a writer turning to nature (inflected by cultural memory) to resolve the stresses of urban social life, a moment displacing a rich awareness of cities-in-nature with the personal vision of a character in need. What gives the moment special significance, though, is how it is gendered. It shows a woman finding a moment of satisfaction in what is traditionally women's work with natural processes, bringing to grow, around the homestead, what nourishes, then tending and harvesting. It is not only her fraught family experience that this moment comments on implicitly but also her sexual experience, as she focuses on the gardens' luxuriant unruliness. The social-sexual experience recounted in "Tropical Fish," the central story in this collection, narrating her sexual encounter with a European male and the pregnancy she aborts, suggests how a woman's most intimate life can be shaped by interconnected patriarchal and colonizing behaviors. Obliquely, her reflective moment supplies a way to contextualize that experience, to give a brief imaginative glimpse of how she might have enjoyed her woman's body, her sexuality, within a society formed by a culture of tending, not domination, one in which women's experience matters. This moment offers intuition of a different kind of relation between the human body's life, with its participation in natural processes, and the social order, a kind of relation that involves a different significance for gender. Gender issues and women's sexuality are more than "social issues," this moment suggests; they have an ecosocial dimension and matter to any focused attention on cities in nature, involving us, as readers and writers, as surely as issues, for example, of water supply or food provision.[18]

As with Duiker's novel, Ngũgĩ's and Baingana's fiction when read fully in the way I propose would draw on an understanding of what supports its cities (Nairobi, Entebbe, Kampala) with their material life and what gives them their social vitality. Such reading would also draw on understanding of the difficulties these cities face. It is in relation to such awareness that the fiction's insights into its social worlds and its characters' experiences, objective and subjective, can be held in creative tension.

CONCLUSION

In closing, I return to general considerations. Urban places involve a rear-ranging of natural processes in accord with social intent—to produce, as much as possible, a beneficial kind of "nature," as inhabitants supply them-selves with food, water, fuel, material for infrastructure and transportation, and so on. Yet nature is not simply inert, subject to human power, and cities have to deal with the unwanted consequences of their growth (pollution, damaged ecosystems, etc.). A globalized urbanism now faces the conse-quences of its long use of fossil fuel as it faces both climate change and also, thanks to the power fossil fuels give to extract and transport resources, de-pleted and acidified oceans, deforestation, etc., and the spread of invasive species (which include, increasingly, bacteria and non-living viruses). Hu-man societies, too, are not fixed, but, either under the pressure of circum-stances or independently, transform themselves—with increasing speed in this globalized urbanism. There, clearly, can be no formula for establishing a successful relation of city to nature, for both society and nature exist in flux, as they strive to flourish. Under these conditions, good governance becomes even more of a social necessity and citizen awareness of what needs to be done, and needs to be done well, essential. Such awareness can only rest on as rich an understanding of circumstance as possible. Literary study can serve as one further means of crystallizing understanding, one that brings to the wider conversation about cities the kind of interpretation that illuminates the subjective forces set in motion by the social pasts, the histories of coloni-zation and resistance that have produced this planet of globalized cities.

For literary texts, no matter their setting, incorporate, draw on, and devel-op attitudes toward the natural world, ranging from recognition of its power to affirming its subordination to human interests to casual reference in pass-ing. In the printed literature emerging from urbanized cultures, "society" and "nature," then, are not in straightforward opposition, but serve to reinforce each other, with specific attitudes toward the urban making plausible specific attitudes toward nature, and vice versa. In many of the world's regions and this is the case in Africa, where the presence of traditional cultures is felt strongly, especially in rural areas, narrative can draw on traditional attitudes toward nature, with all the ethical and political freight that these may carry, to reinforce attitudes toward the city. Hereby, they can help to clarify critical attitudes toward modernity and the global forces that it rests on. Yet, whatev-er insights into attitudes toward cities and nature can be drawn from the study of literary texts, such study needs to be augmented by insights from the study of the objective life of cities, to recall us to all that literature, with its power to move and to influence our imagination, might draw our attention from. In this way, the reading of literature can participate in the project of encourag-ing the kind of awareness our human ancestors needed to have if they were to survive (an awareness that the modern term "ecosocial" attempts to stimulate). Whatever built environment our far ancestors lived in (from nomad's shelter, to settled hut), it encouraged a sense that the cosmos had a social dimension expressive of non-human energies, energies with which the human lived, on which (if they were beneficent) it relied, and among which it could mingle, leaving the ordinarily human behind.[19] The dominant world of airports, wi-fi, gated and guarded luxury communities, of highways and motorized transportation, and so on, develops subjective barriers to entering such ways of being, whether among the wealthy or among the most impoverished, who have to struggle daily to survive, while aware of what they are missing and what defines their life chances. In this world, priorities have shifted, in both objective and subjective dimensions.

Yet, literary study can remind us of currents of need that run counter to modernity's dominant forces. What Duiker's, Mda's, Ngũgĩ's, and Baingana's African narratives all suggest is the need felt within urbanized cultures for a sense of psychological balance that incorporates both social and natural worlds—one that modernity in its restless innovation, in its constant commodification of both social and natural spaces, will continue to undermine. This need to wake to social and natural worlds in an active—interactive—balance, though utopian under modern conditions, does supply a vision to aspire to: a world in which urban places flourish in balance with natural systems, and so provide their members with the objective conditions for psychological balance. How utopian this vision is, the narratives looked at in this chapter underscore and there is no plan of action to be conjured that will bridge the gap in living social time between reality and vision. Yet, in the meantime, there is the ongoing multi-disciplinary conversation about our life in cities that can hold out the hope of better life, a future more optimistic than apocalyptic vision, whether that of the anxiously comfortable or that of the angrily disempowered. What an ecocriticism of the kind outlined here can supply is a tension in the reader's mind between what the literary text urges us to imagine as real and the complex reality of urban existence—as well as (and this is the second site of desirable tension) the utopian goal of a modern civilization that lives well "in place," whether that place is a particular locale or the planet.[20]

NOTES

1. For editorial work, my sincere thanks go to Swarnalatha Rangarajan. I would like also to thank Fiona Moolla for her insightful comments on material in this essay as I was shaping it. For supporting the research, I thank Transylvania University's Kenan Fund for faculty and student enrichment.

2. I resist the label "urban ecocriticism" simply because all ecocriticism, as academic activity, is urban for expressing the concerns of modern civilization. Of course, not all environmental concern—and activism—occurs in urbanized worlds; it can emerge among those on

modernity's margins, with different cultural traditions, who resist modernity's encroachment. The focus on modern cities inevitably draws in histories of colonialism and colonialism's impact on natural systems (no London or Amsterdam—and their "Exchanges"—then no East India Companies). Thinking of ecocriticism as rooted in urbanism does not preclude exploring the textualizing of rural or "wild" places—frequently, in African worlds, in language that predates the impact of colonial modernity.

3. Dirlik offers valuable analysis of the interrelationship of colonialism, modernity, and capitalism. He distinguishes between a "colonial modernity" (evident during the era of formal colonies and Empire) and a "global modernity," dominant in the present postcolonial era. In this latter form, different societies, with different cultural and political histories, express modernity differently and with regional unevenness. Yet, all societies, to different extent, encounter "the dynamic role played in [modernity's] universalization by capitalism" (163). Wanting to keep this sketch brief and focused on these three key terms, I do not work with any of the abundant scholarship on cities. This work, in both social and natural sciences, would "join the conversation," giving detail to our understanding.

4. The focus that DeLoughrey and Handley provide (together with the essays they have collected) draws on a wide range of writing (by Chinua Achebe, Pablo Neruda, and Edward Said, for example); the interpretive frame has value for supplying postcolonial ways of thinking about how a society's apprehension of its living land could enable a resistance to colonial domination. The frame allows a similar way to chart opposition to global modernity.

5. For discussion of "urban metabolism," see Wachsmuth, who notes that Karl Marx first drew the idea of "metabolism" into the social sciences.

6. The most useful model for a materialist understanding of discourse I find in David Harvey's work. Harvey analyzes discourse in relation to five other moments of social process (e.g., institutions) that both validate discourse and are, in turn, validated by it (see *Justice, Nature* [77–95]). I need to acknowledge, as well, a general debt to the work of Raymond Williams (in *The Country and the City*, especially), of John Barrell on British landscape art, and of numerous postcolonial scholars of landscape.

7. I use the term "literature" to include story and song crafted within oral cultures, which those of us reading become aware of, mostly, through the practices of urban cultural transmission (though family memory can still form an important channel).

8. More and more people across Africa experience urban life directly. Rapid urban growth has occurred since the 1950s and urban population, according to a recent study by UN Habitat, is expected to triple between 2010 and 2050. See Schuttenhelm. For an introduction to African cities, see Freund; for an excellent recent study, see Meyers. While my focus on "globalized cities" does appear to privilege Northern frames of reference—and both Myers and Lawhon (addressing how to think about South African environmentalism) explore the problems with such privileging—it suggests a dialectic characteristic of postcolonial thinking. See "Situating Ecology" (298–99), for my comments on a postcolonial interplay of the local and the global.

9. Comments on Mda's novel in this essay rest on a fuller reading in my essay, "Situating Ecology."

10. Shaun Viljoen supplies an excellent introduction to the novel and its characterization of Azure.

11. Viljoen suggests, rightly, that this generic approach to the novel would supply valuable results (xxii).

12. The book and film titles are, of course, strongly suggestive; an up-to-date study exploring the shifting forms of a Euro-American postmodern temporality and its relation to apocalypse is yet to be written. While World War II and its aftermath mark a distinctive moment of origin, the waning of "liberation" cultures in the 1970s, coinciding with worry over "limits to growth," exacerbated by the rise of globalization, post-Cold War—and, before the Cold War ended, the worry over "nuclear winter," and now the fear of terrorism, all suggest the complex interplay of social and planetary collapse that figure in this haunting. Another issue, entirely, is African thinking's relation to these Northern imaginings.

13. It is not coincidental that this orphan grew up in Mshenguville, Soweto—and his age suggests his birth date coincides with Mshenguville's inception, when homeless people invaded a golf course to erect their shacks there. See "About Mshenguville."

14. I acknowledge, here, Julia Martin's account of her classroom practice, one that encourages a re-visioning of the flows at work in a city's life.

15. Humans can survive the winters of Antarctica because they are provisioned from elsewhere; in the globalized world, the natural systems sustaining cities are, indeed, often far from a city's place—but they are there, as are the necessary transportation systems. It is important to note, too, that our "individual" bodies, provisioned within this network, are themselves communities of organisms, including bacteria with which we live in symbiosis. In "our" bodies, bacterial cells outnumber by a factor of ten the human cells. The U.S. Government is conducting research into links between this "human biome" and health. See "Program Snapshot."

16. See Njogu, who writes of the genre's "riddling" dimension, one that Ngũgĩ deploys as he exposes worlds hidden below the surfaces of life in modern Kenya. The novel does not—and cannot—reproduce the full experience of this genre belonging to an oral culture, a genre marked by competitive challenge and response. Moreover, the narrator, in hybrid fashion, identifies as "Prophet" and enacts a prophet's revelation of divine will as he calls Kenya's people to reason together about the story of Warĩĩnga: "The voice of the people is the voice of God" (8).

17. The difference between these representations of concluding violence derives from their narrative's different strategies for representing social reality—which, in turn, can be linked to their different cultural moments. Ngũgĩ, in this novel, commits to interpreting this modern, post-independence Kenya in terms exploring the cultural dimensions of neocolonial theory, though his commitment to Marxist thinking prevents him from resting on a culturalist position.

18. Baingana's collection, while valuing women's solidarity—and satirizing women incapable of it—is quite heteronormative, which this association of gardening and sexuality reinforces. Linking sexuality and nature, though, does not need to reinforce existing social norms. Sexuality in nature, as biology reminds us, is "wildly" various, while gardening, of course, is a form of domesticating nature—"socializing" it and, historically, in ways that reinforce specific familial and social structures. Yet, the binarism is misleading. Gardening need not be seen as restrictive—there is nothing necessary about the social structures it can be associated with. Moreover, it has always involved engaging with "wild" species, creating hybrids.

19. Chinua Achebe's *Arrow of God* illustrates the drama of living within such a world—and encountering a society with a very different understanding of nature and humans—who carry with them an Empire's sense of built environment.

20. This "utopian goal," I argue elsewhere, can be given content in the present by drawing on an expanded sense of environmental justice culture. See "Environmental Justice."

WORKS CITED

"About Mshenguville." Mshenguville Project. Soweto.co.za, n.d. Web. 26 December 2009. http://www.soweto.co.za/html/pro_mshenguville2.htm#about.

Achebe, Chinua. *Arrow of God*. 1964. New York: Doubleday, 1989. Print.

Baingana, Doreen. *Tropical Fish: Stories Out of Entebbe*. Amherst: U of Massachusetts P, 2005. Print.

Barrell, John. *The Dark Side of the Landscape: The Rural Poor in English Painting 1730–1840*. New York: Cambridge UP, 1980. Print.

DeLoughrey, Elizabeth, and George B. Handley. "Introduction: Toward an Aesthetics of the Earth." *Postcolonial Ecologies: Literatures of the Environment*. Ed. Elizabeth DeLoughrey and George B. Handley. New York: Oxford UP, 2011. 3–39. Print.

Diamond, Jared. *Collapse: How Societies Choose to Fail or Succeed*. New York: Penguin, 2005. Print.

Dirlik, Arif. *Global Modernity: Modernity in the Age of Global Capitalism*. Boulder, CO: Paradigm, 2007. Print.

Duiker, K. Sello. *Thirteen Cents*. Cape Town, South Africa: David Philip, 2000. Print.

Freund, Bill. *The African City: A History*. New York: Cambridge UP, 2007. Print.

Harvey, David. *Justice, Nature and the Geography of Difference*. Cambridge, MA: Blackwell, 1996. Print.

"'The Last Word': Sello Duiker." *Words Gone Two Soon: A Tribute to Phaswane Mpe and K. Sello Duiker.* Ed. Mbulelo Vizikhungo Mzamane. Pretoria, South Africa: Umgangatho, 2005. 27–30. Print.

Lawhon, Mary. "Situated, Networked Environmentalisms: A Case for Environmental Theory from the South." *Geography Compass* 7.2 (2013): 128–38. Print.

Martin, Julia. "Situating 'Place' for Environmental Literacy." *English Studies in Africa* 52.2 (2009): 35–49. Print.

McCormack, Ray, and Basil Gelpke. *A Crude Awakening—the Oil Crash.* Docurama, 2007. DVD.

Mda, Zakes. *The Heart of Redness.* Cape Town, South Africa: Oxford UP, 2000. Print.

Miller, John. *Empire and the Animal Body: Violence, Identity and Ecology in Victorian Adventure Fiction.* New York: Anthem, 2012. Print.

Myers, Garth Andrew. *African Cities: Alternative Visions of Urban Theory and Practice.* London: Zed Books, 2011. Print.

Ngũgĩ wa Thiong'o. *Devil on the Cross* (Caitaani Mutharaba-Ini). 1980. Oxford, UK: Heinemann, 1982. Print.

Njogu, Kimani. "On the Polyphonic Nature of the Gicaandi Genre." *African Languages and Cultures* 10.1 (1997): 47-62. Print.

"Program Snapshot." Human Microbiome Project. National Institutes of Health: Office of Strategic Coordination—The Common Fund, n.d. Web. 10 July 2013.

Said, Edward W. "Reflections on Exile." 1984. *Reflections on Exile and Other Essays.* Cambridge, MA: Harvard UP, 2000. 173–86. Print.

Schuttenhelm, Rolf. "Urban Population Growth Africa 300 Percent Between 2010 and 2050." *Bits of Science.* Bitsofscience.org, 24 November 2010. Web. 15 November 2012. http://www.bitsofscience.org/population-growth-africa-cities-568/.

Viljoen, Shaun. "K. Sello Duiker's Thirteen Cents: An Introduction." *Thirteen Cents.* Athens: Ohio UP, 2013. Print.

Vital, Anthony. "Environmental Justice on a Postcolonial Planet? Literature, Home and Dangarembga's Nervous Conditions," in preparation.

———. "Situating Ecology in Recent South African Fiction: J. M. Coetze"'s The Lives of Animals and Zakes Mda's The Heart of Redness." *Journal of Southern African Studies* 31.2 (2005): 297–313. Print.

———. "Toward an African Ecocriticism: Postcolonialism, Ecology and *Life & Times of Michael K.*" *Research in African Literatures* 39.1 Spring 2008: 87–106. Corrected Version: Postcolonial Ecology. Anthony Vital, 2008. Web. 20 July 2008.

Wachsmuth, David. "Three Ecologies: Urban Metabolism and the Society-Nature Opposition." *The Sociological Quarterly* 53 (2012): 506–23. Print.

"Welcome to the Water and Sanitation Website." *Water and Sanitation.* City of Cape Town | iSiseko saseKapa | Stad Kaapstad, n.d. Web. 10 March 2013. http://www.capetown.gov.za/en/water/Pages/default.aspx.

Williams, Raymond. *The Country and the City.* New York: Oxford UP, 1973. Print.

Chapter Fourteen

Environmental and Cultural Entropy in Bozorg Alavi's "Gilemard"

Zahra Parsapoor

INTRODUCTION

In narratives where nature does not play a direct role, it seldom is given importance by critics primarily for itself, and if it is, it is for the reason that nature and the physical environment can be employed in representing the inner world of characters or as an element in the creation of atmosphere. One of the questions Cheryll Glotfelty put forward in her groundbreaking book *The Ecocriticism Reader* is: "What role does the physical setting play in the plot of this novel?" (Glotfelty xix). Without denying the significant role which setting can play in the process of characterization and the creation of other elements of a story, an ecocritic assigns a specific role to nature and the effects it can exert on a work of literature. Whether nature has been used metaphorically or literally, the focus of interpretation shifts away from human characters, thereby calling the readers' attention to nature and its effects on human life. This article attempts to present an ecocritical approach to a famous work of Persian literature titled "Gilemard" by first introducing the readings and interpretations which "Gilemard" has so far attracted, and then by offering an ecocritical angle to show the workings of entropy in different layers of "Gilemard." This inclusive approach engages other theoretical views, such as Marxism and feminism, which are relevant to an ecocritical reading of "Gilemard."

Although the canon of Persian literature began to flourish only after three centuries of silence following the emergence of Islam in Iran, it has developed into a great body of literary works during the last eleven centuries. The history of storytelling in Iran occurred for centuries before the emergence of

Islam, and after Islam these stories were translated into Dari (a Persian dialect). For example, in the *Shahnameh* (The Book of Kings), the national epic of Iran, Hakīm Abu'l-Qāsim Ferdowsī (940–1020) turned some of these stories into poetry, while Nizami Ganjavi (1141–1209) turned the romantic and pre-Islamic story of Khosrow o Shirin into poetry. In Persian didactic and mystical literature there are numerous fables and stories, both in poetry and prose, among which *Maṭnawīye Ma'nawī* (Spiritual Couplets) of Jalāl ad-Dīn Muhammad Rūmī (1207–1273) and Saadi Shirazi's *Gulistan* (The Rose Garden) are notable examples. But such stories do not comply with the modern definition of prose fiction in the form of novels and short stories (Hoghughi 32). In fact, there are substantial discrepancies between Persian stories and modern Western novels and short stories in terms of narrative elements and techniques such as characterization, point of view, and plot. In traditional Persian stories, characters are typical and static, much like characters in medieval morality plays. The setting and events of a story are barely described and the story is set in an unspecified time and place. Characterization is very crude and simple; all speak in an equal tone that does not reflect their social and educational status. Mohammad-Ali Jamālzādeh (1896–1997) is the first writer in Iran who tried his hand at short story writing, using western techniques, and after him Sadegh Hedayat (1903–1951) and Bozorg Alavi (1904–1997) in different ways used and explored the genre. Bozorg Alavi called himself a "realist writer" (The Memory of Bozorg Alavi 50), although romantic and revolutionary ideas are predominantly found in works like *Chashmhayash* (Her Eyes) (1952) and in his writings before 1941 (Mehrvar 79).

"Gilemard" is one of the most famous short stories in Persian literature and is Bozorg Alavi's most exalted short story. Since there is an obvious correlation between the events of the story and the political issues of its time, most of critical responses to "Gilemard" are heavily oriented toward sociopolitical issues and the historical elements of the story. However, I have attempted to approach the work from an ecocritical angle since nature has a significant role in this work.

Born in Tehran to a wealthy merchant family, Agha Bozorg Alavi (1904–1997) was directly involved in the political affairs of his country since his father, Abol Hassan Alavi, took part in the 1906 Persian Constitutional Revolution and his paternal grandfather, Seyyed Mohammad Sarraf, a wealthy banker and merchant, was a leading constitutionalist and member of the first Majles. In 1922, he was sent with his brother to Berlin to study. When he came back to Iran, he taught German in Shiraz and Tehran, where he met and befriended Sadegh Hedayat, a distinguished modern writer, under whose influence he wrote *Chamedan* (The Suitcase) (1934). Later he was involved in the communist meetings held by Dr. Erani, one of the leading figures of the party, and in wake of such activities, Alavi and fifty-two other party members were put in jail in 1937 under the regime of Reza Shah. Following the Allied control of Iran in 1941, a general amnesty was granted, and upon his release he published his Scrap Papers of Prison and Fifty-Three Persons. He continued his political activities and became a founding member of the communist Tudeh Party of Iran, serving as editor of its publication *Mardom* (People). The fruit of his friendship with Hedayat was the new psychological aspect in his writings, and his involvement in leftist politics resulted in stories featuring social issues and idealistic characters. In 1952 he published *Nameh' ha va Dastan'ha-ye digar* (Letters and Other Stories), which included "Gilemard." During the Coup d'état of 1953 which brought down the government of Premier Mossadegh and the ensuing massive arrests and imprisonment, Alavi was in Germany, and as a result he stayed in exile in East Berlin, teaching at Humboldt University until the fall of the Pahlavi dynasty and the emergence of the 1979 Iranian Revolution (Alshokri 260–67).

Alavi's "Gilemard," which orchestrates the plight of oppressed and wronged peasants against the colorful backdrop of nature, lends itself admirably to an ecocritical interpretation. The geographical setting plays an indispensable part in characterization and narration. Although the focal conflict is between peasants and landlords over domination of land, an ecocritical reading can give voice to the ignored land whose voice was stifled under the repressive and relentless rhythm of industrial progress, as is the case in many developing countries. Without creating the necessary cultural backgrounds and prerequisites for changes in mode of production, the flag of industrial development is set on a land which is about to seethe with entropic agitations.

The story begins with the description of stormy weather in the jungles of Gilan with two armed officers escorting Gilemard to Fumanto in order to hand him over at a police station. Gilemard, the son-in-law of Agol Lolmani, the late rebel leader, tries to keep his father-in-law's legacy alive. The first officer, Muhammad Vali, who works for the corrupt government and avaricious landowners, tries to instill fear in Gilemard and other peasants, accusing them of not paying their taxes. The second officer, a native of the hot deserts of Balochestan, comes to Gilan with its extremes of cold and humidity under compulsion of poverty and is a merciless pawn at the hand of masters and easily kills peasants. The captive Gilemard is troubled by thoughts of his deserted son who is all by himself in the cottage. His thoughts about escape are interrupted every now and then by the roaring of the storm and the cries of Soghra, his wife. It is only when they get to a coffeehouse to rest that the first officer tells Gilemard that he has killed Soghra. The second officer, who has his eyes on Gilemard's money, sells him the gun he found in Gilemard's shelter. When Gilemard is alone with Muhammad Vali, he decides to takes his vengeance but changes his mind as Muhammad Vali desperately pleads for his life for the sake of his five children. Gilemard relents and plans to escape wearing Muhammad Vali's overcoat. As he was about to leave coffeehouse for the jungle from where he was hearing the cry of his dead wife, suddenly the second officer shoots Gilemard from behind.

LITERATURE REVIEW

In his article "Recreating the Truth," Mohammad Ali Sepanlo provides a sociological criticism of the work, considering the characters of the stories as representatives of different classes of the society. For example, Gilemard and the Baluch officer are representative of the deprived lower class of the society; the former is aware and conscious of this fact, whereas the latter is ignorant and indifferent. Mohammad Vali is the unwitting victim of a system for which he is sympathetically working (Sepanloo 336).

Jamal Mirsadeghi maintains that the story has both realistic and symbolic dimensions: "apparently Gilemard is realistic and to some extent naturalistic, but the elements of the story are selected in a way that make the symbolic reading of the text possible. In other words, the story in surface is realistic but in depth it has symbolic features" (Mirsadeghi 407). According to this reading, just as the blowing wind, rain, and thunder make the jungle stormy and chaotic, so too do Gilemard's distress and mental strain fill him with an existential horror.

Much in line with Mirsadeghi's view, Zakarya Mehrvar holds that "The atmosphere of Gilemard is replete with horror and nature intensifies such horror, distress, passion, and sense of failure. Wind, rain, jungle, wood, women's crying and wailing give depth to the disaster" (Mehrvar 79). Ezati Parvar sees the jungle symbolically: "Bozorg Alavi also in his works . . . dramatizes the oppressed women with different symbols. If we consider the jungle the symbol of oppression, from its heart the wail of a tortured woman can be heard" (Ezati Parvar 10). Concerning the stormy description of jungle with which Alavi begins the story, Ali Mohammad Honar sees the scenery as foreshadowing what will happened in the story, laying the groundwork for readers to come to terms with ensuing gloomy events (Honar 137).

ENTROPY IN SETTING

I consider setting to be one of the focal points of analysis in ecocriticism, one of the features that distinguish this critical approach from other methodologies. Setting encompasses influencing factors such as time, place, and circumstances which cause the action and take it forward. As Glotfelty observes, "If we agree with Barry Commoner's first law of ecology, 'Everything is connected to everything else'" (Glotfelty xix); we should take into account both natural and the other-than-natural elements of setting since there are unbreakable ties between cultural artifacts and natural ones in which the story occurs and develops. Eco-philosopher Kate Soper talks about the multiple ways in which "nature" figures in environmental discourse:

> employed as a metaphysical concept . . . 'nature' is the concept through which humanity thinks its difference and specificity . . . employed as a realist concept, 'nature' refers to the structures, processes, and causal powers that are constantly operative within the physical world, that provide the objects the study of the natural sciences . . . employed as a 'lay' on 'surface' concept as it is in much everyday literary and theoretical discourse, 'nature' is used in reference to ordinarily observable features of the world: the 'natural' as opposed to the urban or industrial environment . . . animals, domestic and wild, the physical body in space and raw materials. (*The Green Studies Reader* 125)

A clear distinction between these categories is not possible since they are interrelated and influence each other in multiple ways. The manner in which the natural environment can exert a significant influence on cultural and social environments and also on the physical and mental characteristics of inhabitants is an ancient debate going back to Greek philosophers like Plato and Aristotle and to polymaths of the Islamic Golden age like Avicenna and Ibn Khaldoun (Islampor Karimi 41). In *The Treatise of Ikhwanu al Safa* (2007), an encyclopedia encompassing different sciences written anonymously around the tenth century, there is a chapter dedicated to the discussion of the effects of different geographical regions, the height of land, dryness and moisture, the effects of the wind and stars on the human temperament, the corresponding changes in morals and customs and the way of thinking, and so on (151–53).

The separation of nature from culture was erroneously propagated by the Enlightenment. This division was further widened by the first wave of ecocriticism, especially schools of thought like Deep Ecology where, according to some scholars, "in idealizing nature, ecocritics have been guilty of perpetuating the dualistic and alienated distinction between nature and culture generally attributed to Enlightenment" (Becket 38). But the shift toward welding together these two separated spheres has been made by more recent ecocritics with a special emphasis on the dialectical relationship of nature and culture. The interrelation between them does not at all belittle or exclude nature or culture, to the extent that, as Soper aptly pointed out, one should prove the hole in ozone layer exists in nature and not in language (*What is Nature* 124). This interrelationship between nature and culture requires a paradigm shift in our understanding of the accompanying realization that if we are really concerned about the health and survival of culture, we should do something for the health and survival of nature.

"Gilemard" is set in the jungle, which is a natural environment, but the historical events of the Jungle Uprising give a cultural meaning and political significance to this geographical part of nature and also influence it. On the other hand, the jungle with its own specific characteristics largely has influenced the temperament and behavior of characters, coloring the events of story in tangible and palpable ways. Therefore, in "Gilemard" we cannot imagine the jungle without recognizing its cultural factors, nor can the sociocultural environment of the story be properly conceived of without an adequate understanding of the special qualities of the environment of Gilan. The reciprocal relationship between the natural and cultural environment is the question which Alavi has observed both in his characterization and the construction of his plot. "Gilemard" features three kinds of setting:

1. Gilan, with its jungles, which is presented as an undeveloped environment.
2. The deserts of Baluchistan, which are described as having an unfavorable climate with a poor economy and uncultivated culture.
3. The capital city, Tehran, from where the hymns of progress can be heard, features more manmade environments than Gilan and Baluchistan.

These different places have strong bonds with the story's characterization and plot structure. We can see the working of nature in different layers of the story. In the first place, characters' names, indicating identity, are explicitly linked to nature. Two characters have distinct geographic names: Gilemard, which means a man from Gilan, and the Baluch officer, indicating that the man is from Baluchistan. These two regions of Iran are diametrically opposed: one has a rainy, humid, and cold weather and is mainly covered with green trees, while the other is hot and dry, a vast, inhospitable desert. In a similar fashion, Gilemard and the Baluch officer are markedly contrasted. Berg and Dasmann's holistic definition of a "bioregion" as both a "geographical terrain and a terrain of consciousness" (Buell 83) is useful here. The intermingling of the outer and inner terrains can be observed in the attitude and feeling of Gilemard and the Baluch officer. Since the Baluch officer has been sent to the rainy and cold weather of Gilan much against his will, he constantly complains about the unnerving and relentless rain, and longs to return to the hot and dry climate of his native region. The topography of the desert is tied to the Baluch officer's greed. He wants to accumulate money by unethical means and "escape once more to the same hot desert; after all, desert is wide enough to hide him from government officers" ("Gilemard" 399). However, Mohammad Vali is also not in a state of harmony with the nature of Gilan and its culture: "Look like he's going to back of beyond; bringing blanket with yourself. Have you eaten your rice? Hey jowl-eating body. You've better be in Tehran for a while to eat pottage" ("Gilemard" 399). Gilemard reacts to the rain and cold of Gilan in a positive manner since he resonates at a deeper level with the weather of Gilan. The harmonious link between Gilemard and the special climate of Gilan is reflected in the adverse feeling Mohammad Vali bears towards them: "All the unremitting toils he endured along the long way of dark, cold, and rainy weather of autumn, he attributed to Gilemard" ("Gilemard" 399).

In contrast, Gilemard and other farmers have a negative view of Tehran, the epitome of city life and industrial development. They defend themselves against the charge of defaulting taxes by putting the blame squarely on city life and the governmental rules that have brought about chaos and disorder: "You say there's no riot and mess in country? Then what is disorder? You've stripped us, removed us for our homes, what else do we have? There's no peasant left now" ("Gilemard" 404). The peasants brand the government officers coming from Tehran as murderers and thieves: "Why do you get innocent peasants? Why do you kill them? Who is the thief? My great-grandfathers lived on this land. Which of your lords have been in Gilan more than fifty years?" ("Gilemard" 404). The point Gilemard is talking about is very important in proving his claim, since the government officers act like outsiders, disturbing the harmonious rhythm of village life and throwing Gilan into disorder. Therefore, the narrative opposes any kind of change in the environment done in the name of development since it will inevitably lead to cultural opposition and disorder. Tragedy befalls negative characters like Mohammad Vali and the Baluch officers together with their masters, when they are placed in an alien land and culture. The disorder in "Gilemard" can be analyzed in light of entropy both in man-made and natural environment. "Entropy," as a term used by ecocritics, gives a new dimension to the analysis and interpretation of literary works. In his analysis of Poe's "The Masque of Red Death," Hubert Zapf offers a lucid and succinct definition of entropy:

> Entropy—which is a mixtum compositum between the two Greek words ener-geia (energy) and trope (transformation)—is the implication of the Second Law of Thermodynamics that all energy is continually transformed in a way in which disorder increases at the expense of order in the material universe. The available energy will undergo the irresistible and irreversible process of being degraded from useful to less useful forms until . . . maximum entropy is reached which is known as the 'heat death' of the universe. (Zapf 212)

According to this theory, natural processes are generally moving toward maximum entropy and are prone to disorder. The change toward disorder in general is a movement toward annihilation. Entropy, which is first used in physics, now finds its way into cultural and social studies as well. In "Gile-mard" entropy can be observed both in nature and in the sociopolitical and cultural life. The increase of entropy in nature and society is interrelated and has a determining influence on the characters and the events of the story. We can ascribe the existing entropy in the village to the imported elements of city, which stealthily creep in under the banner of development. New ways of farming are introduced which not only change the cultural order of the village but also negatively impact the economic state of the peasant family. Any modern narrative of progress becomes the conduit of the energies of entropy without adequate grounding in culture:

> The phenomena which are deemed as elements of progress making would not necessarily have the same result in every society. But the ramifications of the entrance of such phenomena in societies lacking the sufficient cultural vigor to control these disorders which appear in the form of entropy practically transform into an 'anti-progress' phenomenon. (Rafi'far 57)

The drive toward industrialization derived a major boost in the reign of Reza Shah (1925–1941), who in order to accomplish his numerous projects, ranging from the reformation of the army, construction of roads to the extension of railways, was forced to levy heavy taxes on people, most grindingly on farmers (Ahangaran 155). With the help of government officers, landowners imposed new taxes on peasants, and if the peasants defaulted, their land was confiscated by landowners, and they became obliged to rent their own land from their landowners. In Gilan up to the 1970, more than 90 percent of arable lands were taken by landowners (Ahangaran 159). The emphasis the central government laid on the increase of farm production demanded new farming facilities, which were to be imported into villages to take the place of traditional farming methods: "This process entails the dependence of village life to city and dependence of agriculture to government" (Ahangaran 90). This policy resulted in connecting the city to the village, the dependence of the village and peasants on the city, and the entry of modern socio-cultural elements of urban spaces into village life and the negative reaction of villagers that disturbed the order of village.

Alavi in "Gilemard" depicts the entropy which has permeated the village community. The city, with its myopic policies of progress and attendant cultural manifestations, is pitted against the culture of the village. The plot of the story has been built on this matrix of opposition. We can even see these binary oppositions in the delineation of characters' clothes and their equipment. The character Gilemard is scantily clad in a blanket, which barely prevents him from getting wet, while Mohammad Vali is dressed in raincoat, boots, and cap, protecting him against the rainy weather of Gilan. The Italian pistol of the Baluch officer and Mohammad Vali's gun is contrasted with the stick which Gilemard craves to have. The establishment of a police station by the government against an act which triggers the peasants' protest contrasts with the traditional role of a local sheriff. It is interesting that Gilemard himself is aware of the advantages of elements of city culture with respect to village life. Gilemard complains about the cruelty and treachery of government officers sacking peasants, whereas towards the end of the story he exchanges his blanket with Mohammad Vali's raincoat and cap after having overcome Vali with the aid of a gun. His exchange of clothes with Mohammad Vali is indicative of the fact that the action results from his own free will and was not taken under the force of government, law, or the city. The dialectic exchange between Gilemard (the representative of the unaffected peasant life of Gilan) and Mohammad Vali at this juncture is further accentuated by Gilemard's symbolic act of wrapping Vali in his blanket, thereby erasing his alien presence in the environment of Gilan. These binary oppositions of places and cultures result in the increase of entropy in the social and natural environments of Gilan and also deeply affect the souls and minds of characters, preventing them from thinking and acting wisely. Thus all of their actions are impulsive and unpredictable.

Such entropic disorder can be seen both in Gilemard's family where Gilemard's wife is murdered by Mohammad Vali and the infant is abandoned in the cottage and in Vali's family which has lost its financial stability. Vali holds the peasants responsible for his state of penury: "Then the landlord from where should make his living? How he can pay his taxes? If government has no money, then who should pay us? It was for your action that we haven't been paid for four months" ("Gilemard" 389). A grimmer story can be traced to the Baluch officer who in his childhood witnessed how landlords pillaged his family, killing his father and his two brothers. To save his life he joined the government and was sent to Gilan as an officer. He is a stranger in both an alien environment and alien culture, thus experiencing more deeply the tensions of entropic disorder. The violent coming together of these opposing forces is epitomized in his treacherous action of shooting Gilemard from behind. His ultimate happiness revolves around the puny quest of making enough money to return to the hot deserts where he can find solace. From what has been mentioned we can conclude that the economic dependence of the city to village and village to city leads to the entropic disorder at the political and cultural level of story.

ENTROPY IN NATURE

The connection made by Alavi between the existing entropic disorder in culture and political and the entropic disorder in nature is a point worth mentioning. This simultaneous social and environmental entropy reinforces the linkages between culture and nature. The choice of the jungle as the setting for the story has marked political implications. In this regard, Khanlari notes that "the use of 'jungle' was considered unpleasant, because the use of this word shows that the poet wants to express his affiliation to Mirza Kuchak Khan, the renowned man of forests of Gilan" (qtd. in Alavi, *Tarikh* 211). Kuchak Khan was a rebellious figure who founded the Movement of Forest in Gilan and Mazandaran in 1914 which lasted for seven years. One of the main impetuses behind this movement was the landlords' brutal exploitation of the people (Motavali 75). Therefore, by selecting the jungles of Gilan as the setting of his story, Alavi relates the plot of Gilemard to the sociopolitical context of his time. The jungle is not merely a space where the events of the story take place, nor is it a mere stretch of wilderness. It is the theater where sociopolitical and cultural entropy is played out. The story begins with the description of the entropy in nature:

> It was pouring with rain. The wind was scratching at the ground and was about to wrench it off. Old trees were fighting with one another. From woods the cry of a tortured woman could be heard. The blowing gale had unbounded the silent songs. Chains of rain had fastened the murky sky to muddy ground. Rivers were revolting and the flowing water was pouring everywhere. ("Gilemard" 388)

The natural scene that Alavi depicts is turbulent, as if all elements of nature are revolting against something, which is closely related to the human plot. There are significant correlations between the description of nature with its heavy rain, blowing wind, and revolting rivers and the peasant society of Gilan. In Persian "revolt" is used to describe a river that overflows its banks as a result of heavy rain as well as a society which rebels against the controlling government. With the increase of disorder in nature, we can witness an increase of anger and violence in characters' behavior. However, in the concluding part of the story, Alavi puts forth the view that calmness in the world of nature can also bring about tranquility in the human mind. He first describes the serene and calm atmosphere which now dominates the nature and decreases the entropic disorder; that is the movement toward light and morning and the cessation of storm: "The rain stopped and in the silence and pleasant purity of morning, Mohammad Vali's weakness and timidity filled Gilemard with intense hatred" ("Gilemard" 405). Gilemard wants to take revenge on Mohammad Vali for the murder of his wife and for all the acts of injustice that the Government has wreaked on him; however, the serenity of nature makes him change his mind and he refrains from killing Vali: "Gilemard spat and within a few minutes he took Mohammad Vali's raincoat and bullets, and bound his hands and neck with his blanket. Gilemard put on his cap and raincoat and left the room" ("Gilemard" 405).

The same change of mood also becomes evident in Mohammad Vali's behavior; he loses his overbearing spirit and behaves like a submissive child. The only violent and unruly figure is the Baluch officer who kills Gilemard toward the end of the story. One question can be tackled here: does the increase and decrease in the entropic disorder in nature positively correlate with the increase and decrease in the entropic disorder in the human society? Answering to this question seems to be rooted in the traditional religious and mythological beliefs which are expressed subtly in short stories and novels. We see this especially in epic poems which portray a direct and positive correlation between the increase in violence, injustice, and war and the chaotic and entropic condition of nature. The influence of violence or calmness of a particular environment on the mood and behavior of its citizens is a subject that has attracted the attention of writers as well as sociologists. Like critiques written on stories with similar setting, many critics have interpreted the onslaught of wind, rain, and other natural voices in Gilemard as a mirror representing the cry and uproar of peasants. Hossein Payandeh in his criticism on Alavi's "Lease" presents a similar view of nature. The story begins in a rainy sunset and ends in the rain-washed morning which finds the rented house of Amirza in shambles due to the incessant rain. Payandeh observes that rain is a metaphor for the storm and the fundamental change which is going to ravage all parts of the society, making underlying alterations. Rain is the narrator of a past history, a transition from the feudal order of society to modern capitalism. Resistance against the historical changes is doomed to failure, a futile striving. Thus the timbers of an old house should collapse (Payandeh 418–19). In a similar way Mehrvar makes the wind a symbol of ravaging government officers, and the uprooted old trees are representative of an old nation that is tortured and defeated (Mehrvar 80). This position has been taken by many critics, but if we read closely we find that the violence of nature has been inflicted on the characters, including Gilemard himself:

> Rain didn't want to stop, on and on splashing water into officers' and prisoners' eyes and ears. It was about to wrench the blanket off of Gilemard's neck, taking away the raincoats of the officers. The thunder of thick water smothered the scream of shelducks. Now and then an old tree wrested out of its roots and trembled to the ground. ("Gilemard" 392)

Mirsadeghi also takes the outer storm and rainy and unpleasant weather as a mirror for the tormented mind of Gilemard (Mirsadeghi 408). This mimetic view of nature in which nature is taken as a representative of the inner world of characters' mind has a long tradition in literary criticism. But such view cannot be applied to "Gilemard," since there are some sentences in the text that invalidate this claim and highlight the ambiguity of human-nature relationships. The blowing of wind and storm not only do not help reveal Gilemard's voice, but they actually suppress it: "But [Gilemard's] cry could not reach anywhere because the storm smothered any weak voice in midst of rain and wind" ("Gilemard" 396). In another passage Gilemard wishes the rain would stop: "If the rain stopped, he could have found a stick" ("Gilemard" 390).

From an ecocritical view, we can place the entropy of nature beside sociopolitical entropy or the conflictual disorders in characters, not taking nature as a mirror representing the disorder of a specific class or character. In "Gilemard," it is evident that the entropy of nature is alongside other disorders, signifying that any kind of change made by humans on the physical and cultural environment can lead to entropy, including changes that have been done in the name of progress. The main conflict in "Gilemard" centers on the question of land possession played out between Gilemard, the representative of peasants who are the original inhabitants of the green environment of Gilan, and Mohammad Vali, the representative of government and landlords. The government takes advantage of its power of money and army, whereas the peasants fall back on their ancient culture which gives the right of land possession to those who have worked on it. Each of them has their own reasons for their claim and the reader is free to take side with any of them. However, the ownership of land by humans and the resulting exploitation of it is the emerging vision of the novel. Ecocriticism questions and challenges this axiomatic fact of developing societies, and this is why it is very important to apply an ecocritical lens to literary works like "Gilemard," which focus on the destructive treatment of the natural landscape through the process of economic development. Historical evidence offer us a more telling picture of the story about how the northern lands of Iran, its jungles, pastures, and farms, had been exposed to economic exploitations when Alavi wrote "Gilemard." In pursuit of economic and industrial progress, Reza Khan ordered the railway to be extended and built many factories for producing leather, lump sugar, cotton, silk, cans, glass, and so on. Despite the attraction that such words as "change" and "industrialization" may have for a society, a heavy price is paid by nature. Nature provides the raw material for the factories, and nature receives the waste produced by these factories: "The vast five-year project of agriculture from 1940 in Iran was fulfilled; according to which the production of bean and grain should increase to five hundred thousand tons and such product as tea, sugar, beet, cotton, and linen should increase by fifty to two hundred percent" (Elwell-Sutton 256). Both mechanization and the extension of the land allocated for agricultural production, which lead to an increase of agricultural production, have resulted in

> many damages for nature. Heavy farming machines destroy useful microorganisms of soil. Single cultivation of land and the cultivation of productive grains which is known as green revolution, require heavy irrigation, chemical pesticides and chemical manures which damage the land. The various native plants which are more adapted to the environment and are more resistant to the pesticides are doomed to be destroyed. (Fateme Farmanfarmayian 55–56)

Over-farming, which means the increase in cultivation of land at the expense of destroying it, may be economically justifiable but "it leads to reduction of natural environments and extension of some kinds of plants and animals life" (Goudie 43).

Another prominent voice which articulates the entropic disorder in "Gilemard" is the cry of Gilemard's murdered wife, heard from the heart of the jungle. As with all revolutions which bring about a state in which repressed and silent individuals can speak their mind freely, the "blowing gale" also makes it possible for those "silent songs" to be heard. One of these silent songs is the bitter cry of a tortured woman that is heard amidst the storm. In fact, reader can hear simultaneously the cry of a woman and the rebellious roar of nature throughout the story:

> The blow of wind carried strange screams from the heart of the jungle to the cottage: the wail of a woman, the moo of cows and complaining cries. The more attentive Gilemard tried to hear, the more he could hear, as if the heart-rending wail of Soghra when the bullet hit her belly can also be heard in the midst of storm. ("Gilemard")

A parallel that can be drawn between exploitation and domination of the earth and the domination of women is the crux of ecofeminism. Karen Warren believes that there are more important correlations between the domination of women and domination of nature (Lorentzen 1). Such correlations can be found both in mythology and in the language and mind of different nations. The connection between women and earth has a long history in Persian language and literature. The root of this belief can be traced back to Avesta, the primary collection of sacred texts of Zoroastrianism. *Zan,* which in Avesta means "procreation," is the root from which zam, the land, is taken, and also *zemana*, conception and fertility, derives from this root. Thus *Zamin*, the word which is used in modern Persian language for land, like *zan*, the word which is used in modern Persian language for woman, has at its root the meaning of fertility and procreation (Razi 1239). In Persian poetry, sky is thought to be masculine and land feminine.

> The sky man and the land woman in reason
> Whatever he throws, she fosters (Molavi 256)

"Gilemard" also essentialises the relationship between women and nature. It seems that they are similar in being wronged and suppressed and in their inconspicuous role in society. In most of Alavi's works, women are present, but their presence "does not mean that these works manifest the rights of women according to modern standard" (Mosavi 434). With their presence, they usually give a romantic dimension to his works. In some of Alavi's short stories and in the novel *His Eyes* (1952) women play an active role in political affairs, but the reason for and the way of their presence in such affairs is markedly different from men. It seems that in most of these sociopolitical affairs they are at the service of rebellious men, and they serve as instruments in paving the way for their activities. Even in *His Eyes*, the political activities of the character Faragis are a means to attaining the attention of the painter, Makan, who is desperately in love with her. In his short story "Termite" (1990), the female character of the story, Rughye, is a rebellious figure who lacks the required feminine identity: "The cliché which is used here is that if a woman enters into a more serious matter which has been previously under the reign of men, her action is considered as an act of separating herself from normal life, as if she lacked the feminine identity" (Mosavi 436). Soghra is cast in the role of a catalyst for Gilemard, and it is not clear whether besides being the daughter of Agol Lolman, the leader of rebellions, and the wife of Gilemard, she has any role in the political activities.

The name selected by Alavi for the sole woman character of the story is telling: "The female protagonist of Gilemard is Soghra; Soghra means smaller or minor. It is man who is always greater and major" (Ezati Parvar 10). This nomenclature was used by Alavi's contemporary writer, E'temadzade (also known as Beh Azin), for his female protagonist. In Beh Azin's "The Peasant Girl" the titular character is sacrificed for the desires of landlord's son. The sound of Soghra's wails is mentioned both in the first and last paragraph, and throughout the story her cry reverberates even after nature becomes calm and peaceful.

The sociopolitical history of Iran shows that the influences of these subaltern voices which lack agency, namely those of peasants, women, and nature, have not been effective in changing the sociopolitical conditions. Although peasant protests eventually facilitated the breakdown of the feudal systems and the women's rights movement partially succeeded after many decades of fighting for cultural and economic reforms, the voice of nature continues to remain silent.

Over the past decades, the desire for progress and technological advancement has been considerably motivated in the Middle East. Doubtlessly, all progresses entail some kind of entropy in society and nature. If economic planning can effectively reduce damages caused by entropies in society so too can environmental planning can play down the adverse aftermath of entropy in nature. But in our pursuit of progress, we must not forget that we are in need of natural resources both for advancement as well as for sustenance. Secondly, there is a mutual relationship between nature and society. The effect of the entropy created in each of them will mutually influence the other one. Ecocriticism highlights the mutual relationship between society, culture, economy, politics, and all the matters related to man to nature and his environment. Consilient approaches can help us to consider literary works not merely as a representation of the human emotional and intellectual field but as a conduit for the relational web of the earth where the larger concerns of justice for the subaltern human and greater-than-human others are placed in perspective. An ecocritical interpretation of contemporary Persian literature can reveal the detrimental effects of industrial and technological progress which have been camouflaged by the spiel of politicians and policymakers. Ecocriticism is simultaneously a word of caution and encouragement for progress, a steady and stable progress.

WORKS CITED

Ahangaran, Mohammad Rasul. *Estelahat Eghtesadi Reza Khan Va Ta'hir Avamel Khareji* (Economic Terms of Reza Khan and Foreign Inflluencing Factors). Tehran, Iran: Islamic Revolution Document Center, 2001. Print.

Alavi, Bozorg. *Tarikh Va Tahavol Adabyat Jadide Iran* (The History and Evolution of Iran Modern Literature). Trans. Amir Hossein Shalchi. Tehran, Iran: Negah, 2007. Print.

———. "Gilemard." *Yad Bozorg Alavi* (The Memory of Bozorg Alavi). Ed. Dehbashi, Ali. Tehran, Iran: Thaleth, 2005. 388–406. Print

Alshokri, Fadavi. *Vaghe'garayi Dar Adabyat Dastani Mo'aser* (Realism in Contemporary Prose Fiction). Tehran, Iran: Negah, 2007. Print.

Becket, Fiona, and Terry Gifford, eds. *Culture, Creativity and Environment: New Environmentalist Criticism*. New York: Rodopi, 2007. Print.

Beh Azin, M.A. *The Peasant Girl*. Tehran. Iran: Nil, 1963. Print.

Buell, Lawrence. *The Future of Environmental Criticism: Environmental Crisis and Literary Imagination*. Malden, MA: Blackwell, 2005. Print.

Elwell-Sutton, Laurence Paul. *Reza Shah the Great or New Iran*. Trans. of *Abd al Azim Sabori*. Tehran, Iran: Tabesh, 1956. Print.

Ezati Parvar, Ahmad. "Kand Va Kav Dar Dastan Gilemard, Tajzye Va Tahlile Dastan Gilemard (an Analysis on Gilemard)." *Roshd Zaban va Adabe Farsi* 49 (1998): 10–14. Print.

Fateme Farmanfarmayian, Ahamad Karimi. *Tose'e Eghtesadi Va Masa'ele Zist Mohiti* (Industrial Progress and Enviromental Issues). Tehran, Iran: The Institution of Environment Preservation, 1974. Print.

Gozid Matn Rasa'el Ikhwanu Al Safa Va Khalan Al Vafa (A Selection of the Treatise of Ikhwanu Al Safa and Khalan Al Vafa). Trans. Ali Asghar Halabi. Tehran, Iran: Zavareh, 1981. Print.

Glotfelty, Cheryll, and Harold Fromm, eds. *The Ecocriticism Reader: Landmarks in Literary Ecology*. Athens: U of Georgia P, 1996. Print.

Goudie, Andrew, and Heather Viles. *Earth Transformed: Introduction to the Human Impact on the Environment.* Trans. Rahele Qyasvand. Tehran, Iran: Naghsh Mehr, 2005. Print.

Hoghughi, Mohammad. *Moruri Bar Tarikh Adab Va Adabyat Emroz Iran* (A Survey of Contemporary History of Persian Literature). Vol. 1. Tehran, Iran: Ghatre, 1998. Print.

Honar, Ali Mohammad. "Goroh Rob'e Chegone Shekl Gereft? (How the Group of Rob'e was Formed?)." *Yade Bozorg Alavi* (The Memory of Bozorg Alavi). Ed. Ali Dehbashi. Tehran, Iran: Thaleth, 2005. 121–43. Print.

Islampor Karimi, Askar. *Ta'thir Mohit Zist Salem Bar Ensan Dar Amozeha Islami* (The Effect of Environment on Human in Islamic Teachings). *Pasdar Islam* 300 (2006). 41–46. Print.

Lorentzen, Lois Ann, and Heather Eaton. *Ecofeminism: An Overview. The Forum on Religion and Ecology at Yale.* (2002.) Web. 24 June 2014.

Mehrvar, Zakarya. "Bozorg Alavi, the Revolutionary Romantic." *Yade Bozorg Alavi* Memory of Bozorg Alavi). Ed. Ali Dehbashi. Tehran, Iran: Thaleth, 2005. 75–82. Print.

Mirsadeghi, Jamal. "Tafsir Dastan Gilemard (An Interepretation of Gilemard)." *Yade Bozorg Alavi* (The Memory of Bozorg Alavi). Ed. Ali Dehbashi. Tehran, Iran: Thaleth, 2005. 407–14. Print.

Molavi, Jalal al-Din. Mathnavi Ma'navi. Vol. 2. Tehran, Iran: Amir Kabir Institute, 1994. Print.

Mosavi, Nastaran. "Hovyat Zanan Dar Asare Bozorg Alavi (Women's Identity in Bozorg Alavi's Works)." *Yade Bozorg Alavi* (The Memory of Bozorg Alavi). Ed. Ali Dehbashi. Tehran, Iran: Thaleth, 2005. 428–37. Print.

Motavali, Abdolah. *Barasi Tatbighi Do Nehzate Jangal Va Khyabani* (A Comparative Study of Movements of Jungle and Khyabani). Tehran, Iran: Hozen Honari Sazman Tablighat Eslami Iran, 1994. Print.

Payandeh, Hossein. *Dastan Kotah Dar Iran* (The Short Story in Iran). Vol. 1. Tehran, Iran: Nilufar, 2010. Print.

Rafi'far, Jalal. "Entropy Ejtema'i Va Nagsh Aan Dar Tose'e Farhangi (Social Entropy and Its Function in Cultural Development)." *Name Pajohesh* 1 (1996): 57–68. Print.

Razi, Hashem. *Daneshnam Iran Bastan: Asre Avesta Ta Payan Doran* (Encyclopedia of Ancient Iran: The Era of Avesta). Teharn, Iran: Sokhan, 2002. 37–40. Print.

Sepanloo, Mohammad Ali. "Baz Afarini Vagheeyat (the Recreation of Reality)." *Yade Bozorg Alavi* (The Memory of Bozorg Alavi). Ed. Ali Dehbashi. Teharan, Iran: Thaleth, 2005. Print.

Soper, Kate. *What Is Nature: Culture, Politics and the Non-Human.* London: Blackwell, 1995. Print.

———. "The Idea of Nature." *The Green Studies Reader: From Romanticism to Ecocriticsm.* Ed. Laurence Coupe. London: Routledge, 2000. 123–27. Print.

Zapf, Hubert. "Entropic Imagination in Poe's 'The Masque of the Red Death'." *College Literature* 16.3 (1989): 211–18. Print.

Chapter Fifteen

Environmental Consciousness in Contemporary Pakistani Fiction

Munazza Yaqoob

Contemporary Pakistani fiction in English has a powerful presence in the global English fiction market and enjoys an increasingly wide readership. Politically engaged portrayals of Pakistani culture, incisive critiques of the impact of technology and consumption on human lives, and stark portrayals of toxicity and pollution are nurtured by what Muneeza Shamsie refers to as the "Pakistani imagination" (119). According to Shazia Rahman, theorizing the complex and contradictory realities of today's Pakistan must be located in "the intersections of a geographical and ecological frame" (277). Writers like Uzma Aslam Khan, Mohsin Hamid, Kamila Shamsie, and Muhammad Hanif respond to important environmental issues such as overpopulation and massive food consumption, physical and psychological health of human beings living in urban spaces, value systems of people in relation to natural surroundings, the bond between humans and other forms of life, urban built spaces, and the culture of mindless consumption. In the context of an uneven and rapidly globalizing Pakistan, Ursula Heise's observation about "a sense of planet"—the understanding of how local, cultural practices are now imbricated in the larger "political, economic, technological, social, cultural, and ecological networks"—is key to the understanding of how these networks shape lives on a daily basis in the Global South (55).

As the world's sixth most populous country with a population exceeding one hundred and eighty million people, Pakistan is currently facing a variety of serious environmental issues due to growing industrialization, urbanization, and overpopulation. According to reports of the Pakistan Environment Protection Agency (PAK-EPA), Pakistan suffers from the world's highest environmental pollution rates ("A Brief of Environmental Concerns" 1).

Abysmally low environmental standards for industry have resulted in toxic emissions and water and air pollution that have collectively become the biggest cause of multiple diseases. Pakistan's urban growth is estimated to be 4.6 percent per annum, which is a direct consequence of the influx of rural population into urban centers. Saira Ronaq observes: "Estimates indicate that 6 million (16 percent) people are unemployed and this is expected to increase by 500,000 annually. The increase in population, unemployment, and pressure on agricultural lands means migration to urban areas" ("Environmental Challenges in Pakistan"). This not only contributes to the rapid decay of the urban environment but also has serious implications for the poor whose living spaces are environmentally vulnerable. Although Pakistan's massive urbanization coupled with economics of scale offers opportunities for environmental infrastructure, the nation finds itself unable to use the opportunities. This could be due to the fact that "The cities of Pakistan are currently passing through the stage of risk transition in which modern stresses such as chemicals, heavy metals, and noise combine with the traditional ones such as bacteria and disease vectors" (Khan 508). This contributes to the spectrum of Pakistan's environmental problems ranging from polluted water supply and deforestation to the lack of basic infrastructure and the release of hazardous, man-made chemicals into the soil.

Contemporary Pakistani fiction in English holds a mirror to the appalling environmental conditions in Pakistan, and the writers' environmental consciousness is manifested in their portrayal of growing industrialization and urbanization resulting in the degradation of natural surroundings and resources and the deleterious impact on human habitats and communities. The idea of "nature" does not make sense in these contexts since "It stops being That Thing Over There that surrounds and sustains us. When you think about where your waste goes, your world starts to shrink" (Morton 1). Uzma Aslam Khan's works like *Trespassing* (2006) and *Thinner than Skin* (2012) are representative environmental novels of the times since they highlight numerous forms of pollution such as land, water, and air pollution and overuse of sea resources caused by the ever-growing industrial commerce and expanding urbanization. Similarly, Mohsin Hamid's *How to Get Filthy Rich in Rising Asia* (2013) and Mohammed Hanif's *Our Lady of Alice Bhatti* (2011) provide evidence of how "nature," in the pristine, nurturing sense of the word, has disappeared from overpopulated cities of Pakistan, leading to a contaminated environment marked by people with physical and psychological ailments. These fictional works alert readers to the fact that the scarcity of natural resources has been accepted by urban dwellers as a normal part of their lives in cities. These works document the noxious consequences of so-called human development not only on the external aspects of their society in terms of unsustainable life styles (in the domains of architecture, furnishing, brand-conscious existence, etc.) but also on the human psyche and moral values and norms. Novels like *Broken Verses* (2005), *Moth Smoke* (2000), *Our Lady of Alice Bhatti* (2011), and *How to Get Filthy Rich in Rising Asia* (2013) depict in detail the absence of nature in human affairs and the resulting structural changes in human lifestyle and relations. These novels delve into emotionally barren inner lives, psychologically disturbed behaviors, polluted and diseased minds, the materially-driven goals of their characters, and also the cravings of modern city dwellers for peace in their inner and outer worlds. Novels like *Trespassing, Thinner than Skin,* and *How to Get Filthy Rich in Rising Asia* demonstrate how unequal development can impact human domination along the lines of environment, gender, and ethnicity. Hamid refers to the adverse effects of the rampant global capitalism, which is rapidly destroying the indigenous cultures and ecological systems of Pakistan. Uzma Aslam Khan writes in defense of the marginalized communities with the objective of attaining social justice for all and challenges hierarchical dichotomies not only between humans and nature but also between men and women and the powerful and the weak.

The setting of Hamid's novel is Lahore, whereas that of Khan's, Hanif's, and Shamsie's is Karachi. Lahore and Karachi are the most developed cities of Pakistan, and the novels of these authors highlight the detrimental impact of development and modernization on these two cities. The novel *Broken Verses* is set in modern-day Karachi with its mechanized and commodified lifestyle in the foreground and the slow and calm sea somewhere remotely in the background. The entire setting is overshadowed by the modernized and fast-paced life of the elite of this metropolitan city, apparently luxurious and urban, but inwardly barren, hollow, disfigured, and discontented. Mohsin Hamid's *Moth Smoke* is set in another urban centre of Pakistan, Lahore, where industrialization and urbanization have suppressed both nature and humans, driving the latter to depression and frustration. The setting of Mohammad Hanif's narrative *Our Lady of Alice Bhatti* is again Karachi. The world presented is bleak, sick, and grim, where humans are subject to physical ailments and diseases, cognitive distortions, and psychological disorders. Thus, the fiction of almost all of these Pakistani authors is woven around modern-day Pakistani urban life and documents the intricate details of their altered life patterns.

Set in the chief metropolitan cities of Pakistan, the fiction of these authors depicts the surroundings in detail, from the dust-laden air and the smoky and murky sky to the narrow and cramped spaces people inhabit. For instance, Hamid in *Moth Smoke* draws a vivid picture of the falling of trees and other changes coming into Lahore on account of modernization. In *Broken Verses*, too, there are references to the industrialization of Karachi, a city which once abounded with nature but is now reduced to a heap of skyscrapers, smoky skies, dusty roads, and traffic snarls. Hence these texts record the diminishment of cityscapes and naturescapes in wake of myopic progress and development.

Another significant feature of Pakistani environmental fiction is its frequent allusions to the narrow and constricting places that modern urban lifestyle offers. Modern man, as depicted in Shamsie's *Broken Verses*, is asphyxiated on account of the constricted spaces he inhabits which affect both his body and spirit. Modern architecture in this context is devoid of the brightness and openness of space and is detached from the natural surroundings. Aasmani, the central character in the novel, feels suffocated on account of the cramped conditions in which she lives. From her flat and her office nature is visible only in distant, small patches through a window or from a balcony. Her miniscule office has no window at all; however, she is delighted to find that it has a vent. Shamsie in *Broken Verses* depicts the fast-paced lives of the inhabitants of this metropolis who are engrossed in the vicious circle of their market-driven lives, isolated and incarcerated in their own shells, devoid of essential human happiness. Her colleagues at STD studio, where Aasmani works, are all engrossed in hard work to make the studio big and successful. This presents a microcosm of the mechanical life of city dwellers whose entire existence revolves around material gains.

The authors deeply engage with environmental issues like pollution and its multiple manifestations in the metropolitan cities of Pakistan. *Moth Smoke* presents hovering images of dust, smoke, stench, and heat. The atmosphere remains grim throughout whether the scenes are set in lawn parties, air-conditioned rooms, workshops, or streets. There are numerous instances where the lack or absence of nature is mentioned such as "there are no stars because of the dust" (16) and everything is "dulled by a layer of dust" (95), and "the sun is completely blotted out by a dirty sky" (99). The novel presents the readers a world that is decaying and disintegrating fast. Hamid's novel is punctuated with frequent references to heaps of garbage and filth scattered in the surroundings. *Broken Verses* also focuses on a similar theme. Throughout the novel there are references to human interactions; however, the relational web is not functional since it is overshadowed by noise and pollution. While eating in a restaurant, Aasmani observes, "From the other side of the restaurant's low boundary wall, came the sound of the trucks traversing the highway" (266). In *Our Lady of Alice Bhatti*, Hanif also draws attention to the people living in unhygienic and claustrophobic built spaces, which do not accommodate terms like "standard of living" and "quality of life." The atmosphere of these spaces is defined by "a mixture of humidity and sweat" (174), including fungus on the crockery they use and the water taps from where they get their drinking water (104). Therefore, the human bodies reduced to free floating debris in the hinterlands of a society driven by the demands of an exploitative corporate globalized capitalism are, as Hanif comments, "miracles of malnourishment" (93).

Pakistani fiction critiques the impact of global capitalism in developing countries such as Pakistan. These authors are sensitive to the resulting degeneration of modern humans into mechanized beings who are bereft of true freedom and agency since they unquestioningly submit themselves to the demands of global capitalism. These writers regard the endless pursuit of the so-called high standard of living, at the cost of loss of contact with nature and concern for ecosphere, as morally corrupting behavior. In *Our Lady of Alice Bhatti*, Hanif engages in detailed portraits of cognitively distorted people living in these polluted conditions who desperately seek comfort in sex, crimes, drugs, and corruption. They have no bond with non-human nature since their bonds with fellow humans have sickened and weakened. These people being spiritually sick and deprived of genuine happiness seem to be occupied in endless efforts of generating violence and misery in the lives of others around them. This novel offers a panoramic view of Pakistani society with its economic and social injustices, marginalization, and multiple forms of oppression and lawlessness.

Similarly, in Hamid's *Moth Smoke*, society is marked by monetized hierarchies where most of the characters intensely reflect this capitalist thinking and are caught in the frenzied, self-centered spirals of chasing wealth and luxury. Morals do not define people in this capitalist society since respect for the other is based on superficial yardsticks like social status and standing. Relationships in the society are not social relations but market relations. The indulgence of the central characters Mumtaz and Daru in smoking, drinking, drugs, and sex is in fact a reckless attempt on their part to escape from their present conditions of desperation and dissatisfaction with their lives. This moral corruption and degeneration can be seen in almost all the characters of different economic classes in the novel such as Daru, Mumtaz, Ozi, and Murad Badshah.

Mohsin Hamid demonstrates the complexity of mosaics in urban ecology by placing his characters in polluted spaces and tracing the connection between ecological conditions and human behavior. Ozi, Mumtaz's husband, flouts his privilege by recklessly exploiting the material resources available to him. Hamid specially refers to his love for air conditioning in summer and an over-heated room in winter. Ozi has been portrayed as an arrogant man who has developed insensitivity toward human relationships due to his excessive absorption in material comforts and luxuries. He ruthlessly kills an innocent child and frames his friend Daru for the murder. Mumtaz, his wife, feels suffocated by Ozi's extravagant parties and his excessively air-conditioned bedroom and finds her marriage barren and unsatisfactory. She therefore seeks solace in an illicit relationship with her husband's best friend, Daru, who works as an undercover reporter writing about prostitutes under a pseudonym. She tries to find a panacea for her disturbed emotions by writing a critique of the unjust and corrupt practices of her society.

In *Moth Smoke* Daru and his family represent urban middle class people who do not own industry or agricultural land but work in industries, banks, and other professional institutions which are established and controlled by the industrial class. Therefore, their economic resources help them to sustain themselves but they cannot afford material luxuries and extravagant life-styles. Daru, like the rest of the middle class, lives in a claustrophobic house which is steeped in darkness due to shortage of energy resources and other material resources. The pervasiveness of socio-economic injustice makes Daru a dissatisfied person, and he tries to escape his limiting environment through drugs and illicit sex. The tragedy of Daru is actually the tragedy of urban middle class, which is exploited by industrial class. The working class in the novel is represented by Murad Badshah, who in spite of having a Master's degree in English is unable to find employment. As a result, he becomes a rickshaw driver and goes on to become the owner of many rick-shaws and finally ends up becoming a drug dealer and a criminal. Hamid's account of Murad Badshah and Daru sums up the unjust social and economic policies in Pakistan, which compel intelligent and hardworking people to indulge in such corrupt practices. The author equates the life of these two characters with the spaces of their habitation: dark, polluted, dusty, and un-hygienic. Their sweaty bodies are appropriately related to their corrupt actions and settled in their polluted spaces.

Broken Verses tells a similar tale of urban people. The central character, the thirty-one-year-old Aasmani Inqalab, who lives in a flat in a multi-storied building, is dissatisfied with her life and is unable to free herself of the continuing chronic depression and boredom in her life. Aasmani, like many other characters in the novel, such as Ed and Shehnaz Saeed, passively goes through the business of living without any real sense of belonging.

The central characters in *Moth Smoke*, *Broken Verses*, and *Our Lady of Alice Bhatti* desperately try to fake happiness through the mindless acquisition of designer goods, furnished homes, brand new cars, and other accessories and luxuries. The narratives inform us that they are inwardly utterly isolated and paranoid and, despite seeking refuge in materialistic ambitions, false-relationships, sex, drugs, and drinking, enjoy neither peace nor happiness. Although the three novels do not present stories of people whose experiences are informed by nature and deep ties with non-human forms of life, they point in a very subtle manner to the profound interconnections that bind together human and all other-than-human forms of life. These novels, in their concern for urban ecology, are representative of the nature/culture binary-dissolving writing that is typical of the second-wave ecocriticism which moved away from first-wave concerns which set nature apart from large human population and focused on issues like metropolitanism, industry, and the impact of technology on the environment. The redefinition of the word "environment" now came to include urban spaces and the cultures emerging from these spaces as new, interdisciplinary fields like "urban ecology" explored how humans and ecological processes can co-exist in a sustainable fashion in human-dominated environments. Second-wave ecocriticism took up for discussion issues pertinent to urban ecology such as toxic waste, disposal of waste, public health, and environmental justice issues relating to humans who do not have the choice or the social and economic power not to live in decaying urban sites that are marked by environmental degradation.

Uzma Aslam Khan's *Trespassing* and Hamid's *How to Get Filthy in Rising Asia* offer striking examples of urban environmental fiction. *How to get Filthy Rich in Rising Asia* not only focuses on technology's effect on Pakistan metropolitan society but more importantly it draws attention on the impact of global business corporate and capitalism on the indigenous postcolonial societies. Hamid shows how the globalized corporate has intruded into and devastated not only the urban but also the rural biosphere of the postcolonial societies as they have become the storehouses for the developed worlds' industrial toxic wastes. Hamid criticizes this capitalist intervention, as it has done irreversible damage to the various life forms both human as well as non-human in the postcolonial world.

How to Get Filthy Rich in Rising Asia is set in Lahore. Although a small part of the novel is set in countryside, the major part of the novel deals with the urban setting. The disparities between city and village life lie at the center of the novel. One significant aspect highlighted by the author is the devastation caused by industrialization in rural settings as they have become places to dispose industrial waste. As the setting moves from the country to the city, dirty streets are replaced by the "paved ones," potholes grow less frequent and then disappear, electricity appears, there's more dense traffic, and well-constructed concrete, multi-storied buildings (13–14). So nature and natural resources are harnessed to attain the so-called developed standard of living. In the novel, lavish buildings and houses, furnished offices, quick elevators, and luxurious hotels with their barred windows and high walls symbolize places of incarceration.

Water pollution is one of the central concerns of the novel. Starting from the village where there is no clean water to drink, the narrative takes us to the slums where the cracked water pipes are responsible for the mingling of water from the sewers with drinking water. The writer further exposes how the water available in market that is marketed as purified, filtered, and clean is also impure and polluted. For the author one of the main causes of the water and air pollution is overpopulation. Water is the most affected resource in developing and emerging countries since the hydrogeological setting of the region undergoes extensive deterioration due to the urbanization process. The future of the modern urban cities is predicted in the novel where the availability of the drinking water is made a luxury available only for a certain class. Therefore one of the unique services provided by a housing society is drinkable water. The rarity of the water and other resources is not only the result of excessive consumption and pollution but also of negligence of the authorities who are ignorant of the "countless leaky pipes and seepy, unlined channels" (205).

Hamid also documents the adverse effects of globalized business corporate on human affairs and relations. The modern urban man in Pakistan, completely devoid of eco-consciousness, considers the metropolis as an eco-system where only humans exist. So we see that in this battle characters end up leading programmed mechanical lives, "merely going through the motions of . . . life, of rising, shaving, bathing, dressing, coming in to work, attending meetings, taking phone calls, returning . . . for no real purpose" (182). Characters do not aspire for lofty ideals or an authentic way of life since they focus excessively on the need for material gains. Corruption in its varied forms is common in the society and considered to be essential for going up the social ladder. The metropolitan dwellers in Hamid's text are engaged in endless pursuit of material luxuries and achievements and are unconscious of the dynamic role that non-human life forms play in the eco-system. Their knowledge of the biotic community is confined to zoos, to animals imprisoned and tamed in order to provide entertainment to human beings. Nature encased, controlled, and tamed through technology becomes the sole source of urban man's distorted knowledge of the biosphere.

Thus the critique of the growing influence of the global corporate world and technological development on the postcolonial cultures is at the center of Hamid's recent work *How to Get Filthy Rich in Rising Asia*. Besides Hamid, the Pakistani novelist whose ecocritical concerns are most pronounced and advanced is Uzma Aslam Khan. Her works *Trespassing* and *Thinner than Skin* can be studied in the context of postcolonial environmentalism and ecofeminism. These novels not only criticize the built and urban spaces and the intrusion of industry into the natural landscape but also extend their concern to the marginalized communities of the postcolonial cultures such as Pakistan. The author passionately advocates the cause of social justice movements and emphasizes the interconnectedness of the ecological web that binds life forms while discussing issues of environmental degradation. She manifests her genuine concern for the damage done to indigenous cultures and societies on account of this forced intrusion of the globalized market into the contexts of Third World ecosystems.

Khan's *Trespassing* is set in both Karachi city as well as the coastlines of the Indus Valley. The narrative opens with a description of the human invasion of spaces owned by non-human species. Humans are seen as a threat to non-human habitations and the narrative also directs the reader's attention to the encroachment of the habitats of indigenous groups (Indus Valley fishermen) by industrial corporates, resulting in polluted and fissured landscapes. Khan weaves into the narrative the nostalgic lament of the turtle who is digging her nest alongside industrial encroachment: "how much safer it had been when the coastline belonged to the fishermen" (1). The turtle, which is terrified by the coastline lights and intruding crowd of human species, is unable to carry out her natural lifestyle. The turtle's trajectory is mirrored in the displacement and forced migration of the indigenous people, who are forced to abandon their traditional way of life as a result of which they are alienated from their habitat and means of livelihood. In the opening chapter humans who torture the sea turtle and destroy its eggs allegorize the corporate giants who overuse natural resources and displace the poor, indigenous communities. *Trespassing* provides a number of references to highlight how multinational companies are depleting natural resources (2, 122). The coastline and sea resources no longer belong to the fishermen who are the indigenous inhabitants of the Indus Valley. The novel provides significant insight into how the corporations, with their power of technology and capital, have trespassed the boundaries of business and come to control the sea, the fishes, and even the lives and finances of the fishermen (2). The author mourns the lost beauty and grandeur of the Indus Valley which has dwindled into a deserted and "parched" landscape (101) and the loss of wealth of natural resources, which in earlier times had provided daily sustenance to the inhabitants (124). The novel depicts how the massive business activity generated noxious smoke and the use of granite and other chemicals for fishing contributed in making the environment hazardous for human living (126). Foreign companies were also issued licenses by the state to carry on their business without making any arrangements for the local inhabitants (236). These, for the author, are the hard conditions brought about by the international business that have driven the regional inhabitants out of the valley into the plains, where they are regarded as intruders.

Trespassing also narrates the story of the Chinese princess who discovered the silk produced by silk worms and prepared a robe for the emperor by experimenting with technology. The author through this story also traces the history of silk production and refers to Persian smugglers and traders who sacrificed their lives to obtain the silk and worms in order to build their silk industry (11). Here the author attempts to draw the attention of her readers to the history of human killing and wars for controlling and collecting natural recourses. This also tells us the story of human greed and desire to harness nature for the money-making business. There are many references to the exploitation and manipulation of natural resources by man in *Trespassing*. Uzma Aslam Khan voices her criticism of the human greed for the possession and control of natural resources through Dia, one of the characters in the novel who while watching silkworms in her mother's factory comments about how the Empress's actions shape the destiny of others: "If she had known that a thousand years later, several dozen Persians would pay with their lives for trying to smuggle silkworm out of China, would she have made that robe? If she hadn't, perhaps one of the many innocent daughters of those murdered men might have one day stood the chance of discovering something else (11). However, the author seems to support industrial projects undertaken by the native Pakistanis, since she believes that natives who are raised on the land are more ecologically responsible than foreign investors and industrialists. The novel weaves into its narrative tidbits about ecological sensitivity like Riffat's successful experiment of planting a mulberry forest in Thatha, which avoids the use of chemical pesticides, and her silk industry, which employs indigenous methods of organic dyeing.

The narrative of *Trespassing* both acknowledges and privileges the presence of nature and non-human life in human settings. Nature does not merely serve as the background or as the provider of symbols and images to illustrate human affairs, but nature and non-human creatures form a consistent part of the text. The presence of silkworms and their growth process constitute a very significant dimension of the novel. Khan describes in detail how industry and technology have hampered natural processes and disturbed biotic integrity. Khan informs her readers that silkworms kept in factory breeding cells have forgotten how to eat. Some workers had to "chop up their food in tiny slivers and change the supply nine times daily or the fussy creatures would starve" (105) We have an account of nature that is out of step with its own rhythms and is dependent on support from technology and man-made apparatus in order to survive in the harsh environment of commercialized projects. The novel also brings out the contrast between traditional farming and commercial corporate farming supported by technology and use of pesticides.

Like Hamid, Khan also draws attention to the all-pervading lack of hygiene in urban spaces by employing terms like "the grubby halls of the hospital," "dust-opaque windows," smoke of "burned litter," and "the noxious fumes" (69). Despite claims to modernization and an apparent rise in the standard of living, the novel portrays Karachi as a degenerating city since sickness, drugs, crimes, and perverted sexual relations form an integral part of the metropolitan lifestyle. The urban spaces especially metropolitan cities like Karachi are over populated due to massive migration from Indus Valley and other places in the countries, and all are concrete with hardly any greenery around (56). The metropolis has become subject to all types of pollution such as noise, smoke, filth, heat, and dust. This pollution is caused by vehicles, shanty towns, open gutters, garbage and trash piles, industrial waste, and chemical smoke. The metropolis has an acute shortage of resources such as water and electricity (327).

The study of *Trespassing* in the context of social ecology draws our attention to the details about the displacement and dispossession of indigenous communities of Indus Valley during the process of extension of Karachi. Salaamat, the displaced fisher boy, observes that the native modes of existence in his village were replaced by urban methods introduced by corporate industry eventually leading to the loss of their freedom. Salaamat is representative of the people belonging to the rural, natural environment who are compelled to leave their indigenous lands and in the process lose their identity and become non-entities in the urban settings. For instance, Salaamat is called an "ajnabi," "alien," and "outsider" (131). It comes to him as a painful realization that his people the "original inhabitants" of Karachi, have been "pushed to the periphery, and the native populations forced to work under outsiders who claimed the city belonged to them" (132). Khan condemns this intrusion and powerfully pleads the case of the local masses, which become victims of this invasion by foreign businesses and industrialists and are denied their essential human rights to identity and freedom.

However, the novel which can be truly labeled as an "environmentalist text" in Uzma Aslam Khan's oeuvre is *Thinner than Skin* (2012). Khan articulates some of the core issues of postcolonial environmental concerns in this novel. Like much postmodern and postcolonial fiction, *Thinner than Skin* merges "fictional and factual histories and geographies" (Fletcher 4). The main focus of the novel is on the role of politics of war, state and state institutions, transnational tourism, and commerce in bringing about the ecological crisis. The novel attempts to bring "Leopold's Land Ethics" to public consciousness (Mallory 67) and becomes a strong voice for the conservation of the valleys in the northern region of Pakistan. Khan's own land ethics, as presented in the novel, foreground ecological values like living in harmony with nature and other non-human actors.

The novel presents two stories which run parallel; one story revolves around transnational human figures, American and Pakistani-American tourists who travel to the northern regions of Pakistan to photograph the beauty of the northern area, and the second parallel story is of nomads of northern region who are the indigenous inhabitants of the region. The setting of the narrative is the northern region of Pakistan, which is known for its mythical natural beauty and nomadic lifestyle. Through this parallel pattern the novelist juxtaposes the lives of urban people and the indigenous community in order to compare and contrast the mechanistic and artificial life style of urban people with the holistic and sustainable lifestyle of indigenous people. Through the character of Nadir, the photography enthusiast, Khan points to the attitude of pervasive instrumentalization whether regarding the woman's body or the earthbody of mountains, glaciers, and grasslands. To Nadir these objects are beautiful and meaningful as long as he can perceive them through his camera lens and thus remains slave to image rather than reality. Similarly he finds happiness not in the true relationship of love rather in sex. In contrast to this nomad, Maryam, her children, and others live their lives connected to other human beings, animals, and the moods of the seasons and the valley. The life of these nomads is marked by the absence of technical gadgets and artificial pleasures related to them. The novel incorporates a significant discussion on international war politics and state-oriented business plans, which plunge the lives of nomads and their natural surroundings into a crisis. For the indigenous Gujjar tribes of Pakistan, man-made maps are of little significance since they recognize only the bioregional markers of mountains, steppes, deserts, and oases. Since the Gujjars do not conform to state-defined borders they are perceived as outsiders by the government.

The novel offers striking instances of the shrinking of the commons. Dispossession and deepening poverty levels ensue as a result of the blatant negligence of commons stewardship which manifests as fines imposed by the government on pasturelands that have been the traditional grazing grounds of tribals and in the practices of intimidation employed by forestry inspectors to exploit the tribals. The novel also holds up for scrutiny the misleading fictions of grand narratives of democratic capitalism that promise to endless growth and the wellbeing of all. It critiques the new development schemes that play havoc by replacing local animals with foreign species in order to increase international trade and earn profit. This unthinking intervention by the state places severe restrictions on the freedom of nomad tribes, which acutely cripples their "wandering life style" (189). The author explains how the decisions of the government to replace the Kaghan and Kilan goats and "desi," or local sheep, with Australian resulted in economic crisis for the family. The Australian sheep alien to the land could not survive sudden "snow drifts" and "icy winds" (190) and "ate all the food and left the indigenous goats bleating in hunger" (191).

The author moves on and widens the canvas of her narrative to incorporate incidents related to drone attacks, military training camps and vehicles, terrorist groups, and bombing in the valleys along with the beautiful descriptions of giant glaciers, fast flowing rivers, snow peaked mountains of Nanga Parbat, and the myths of the valleys. The narrative thus highlights how military and terrorist groups have put to risk the beauty of the valleys and traditional lifestyles rooted in the ancient history and ecological balance. The author informs us about the militants who have devoted themselves to holy war against Sunnis in the valley (254) and the police and military who are engaged in search of these militants. For both of these groups, the valleys and forests have no significance except for using them for setting up training camps. The author refers to Mariam and her tribe as those whose lives have been "taken over" (299) by these groups and who could feel that the peace was disappearing from the valley. Thus Khan highlights the tragic predicament of the northern landscapes of Pakistan on account of increasing militant encroachments.

Contemporary Pakistani fiction in English not only illustrates the process of growing urbanization and life in metropolitan cities but also reflects on and critiques the reshaping and remaking of Pakistani lives and culture under

the influence of growing consumerism, technology, and capitalism, and re-
sultant migration of indigenous communities to urban spaces. The metropolitan centers as represented in these novels display the growing impact of
industrial development in the form of deterioration of the natural environment and marked decline and degeneration of human society and culture.
The environmental concerns of Pakistani authors pose a challenge to the
grand narrative of technology-fueled progress since their fiction participates
in the global movement for the preservation of nature. Thus Pakistani fiction
in English has become a site of resistance against imperialistic policies of
globalized commerce and industry as the authors have forged a strong link
between the agenda of social justice in postcolonial societies and ecological
justice. These writers who respond to social challenges and make a construc-
tive contribution to society can be best described using Toynbee's term:
"creative minorities" (259). Their voices, though not large in number, are
genuine and powerful as a result of which their presence is felt in their own
society as well as the global arena. Their role is pivotal in raising environmental consciousness for sustainable development in a developing country
such as Pakistan. These authors present an informed critique of industrial and
commercial development and projects of corporate commerce and show their
concern for the reshaping of environment in the urban spaces of Third World
countries. Their novels, which provide an intimate understanding of the urban ecology of contemporary Pakistani society, can be regarded as emphatic
voices from the Global South as they perform the much-needed task of
environmental sensitivity and stewardship ethics in the contemporary world.

WORKS CITED

"A Brief of Environmental Concerns–Pakistan Scenario." Web. 11 July. 2014.

Fletcher, Lisa. "Reading the Postcolonial Island in Amitav Ghosh's *The Hungry Tide*." *Island Studies Journal* 6.1 (2011): 3–16. Print.

Heise, Ursula K. *Sense of Place and Sense of Planet: The Environmental Imagination of the Global*. New York: Oxford UP, 2008. Print.

Hamid, Mohsin. *How to Get Filthy Rich in Rising Asia*. New York: Penguin, 2013. Print.

———. *Moth Smoke*. London: Granta, 2000. Print.

Hanif, Mohammad. *Our Lady of Alice Bhatti*. Noida, UP, India: Random House 2011. Print.

Khan, Muhammad Aslam. "Problems and Prospects of Urban Environmental Management in Pakistan." *The Pakistan Development Review* 35.4 (Winter 1996): 507–23. Print.

Khan, Uzma Aslam. *Thinner than Skin*. Noida, UP, India: HarperCollins, 2012. Print.

———. *Trespassing*. Islamabad: Alhamra, 2006. Print.

Mallory, Chaone. "Acts of Objectification and the Repudiation of Dominence: Leopold, Eco-feminism, and the Ecological Narrative." *Ethics and the Environment* 6.2 (2001): 59–89. Print.

Morton, Timothy. *Ecology Without Nature: Rethinking Environmental Aesthetics*. Cambridge, MA: Harvard UP, 2007. Print.

Rahman, Shazia. "Karachi, Turtles and the Materiality of Place: Pakistani Ecocosmopolitanism in Usma Aslam Khan's *Trespassing*." *Interdisciplinary Studies in Literature and Environment* 18.2 (Spring 2011): 261–82. Print.

Ronaq, Saira. "Environmental Challenges in Pakistan." *Sharnoff's Global Views* 28 February 2014. Web. 29 July 2014.

Shamsie, Kamila. *Broken Verses*. London: Bloomsbury, 2005. Print.

Shamsie, Muneeza. "Beyond Geography: Literature, Politics and Violence in Pakistan." *Journal of Postcolonial Writing* 47.2 (May 2011): 119–254. Print.

Toynbee, J. Arnold. *A Study of History*. Abridgement of Volumes I–VI. Ed. D.C. Somervell. London: Oxford UP, 1946. 230–40. Print.

Index

About the Contributors

Benay Blend is an adjunct professor in Native American, American, and New Mexico history at Central New Mexico Community College in Albuquerque. Her major areas of interest include Western women writers, environmental issues, and the history of the American West. She contributed an article on Chicana, Native American, and women writers of the American West to the 1998 volume *Literature of Nature: An International Sourcebook*, edited by Patrick D. Murphy. She has published articles on Linda Hogan and Gloria Anzaldúa, and her current research focuses on Latin American Jewish women writers.

Charles Dawson is founding New Zealand vice-president of the Australasian ecocritical network ASLEC-ANZ and an associate editor of the journal *ENNZ: Environment and Nature in New Zealand*. He holds a doctorate from the University of British Columbia, Canada ("Writing the Memory of Rivers"). He has worked in Treaty of Waitangi policy and Waitangi Tribunal claims facilitation and is presently Co-Director of a New Zealand study-abroad program for Minnesota's social justice and sustainability education consortium HECUA. His reviews, poetry, and articles have been published in Australasia and North America. He credits the Atihaunui-A-Paparangi iwi (tribe) of Aotearoa/New Zealand as his main river teachers, complemented by many fine works from the United States and Canada.

Sharae Deckard is a lecturer in World Literature at University College Dublin. Her research interests include postcolonial eco-criticism, particularly in South Asian and East African writing, world-ecology, and world literature. Her monograph, *Paradise Discourse, Imperialism and Globalization: Exploited Edens,* was published in 2010. She has edited a special issue of *Green Letters*: *Studies in Ecocriticism* on "Global and Postcolonial Ecologies" (2012) and is co-editor of a special issue of *Journal of Postcolonial Writing* on "Postcolonial Studies and World Literature" (December 2012). She recently published an article on Sri Lankan eco-critical writing, Romesh Gunesekera and John Still in the collection *Postcolonial Green* and has articles forthcoming on Rana Dasgupta and world-ecology (*Eco-gothic*, Manchester UP); Lindsey Collen and the Indian Ocean (*Interventions)*, and Caribbean storm ecologies (Palgrave), among others.

Christopher Lloyd De Shield is a Belizean academic currently residing in Malaysia, where he is a PhD candidate at the University of Malaya, focusing on ecocritical postcolonialism and literary comparison. He is also an adjunct lecturer at Open University Malaysia, where he teaches courses in World Literature, Postcolonial Writings, and Philosophy of Literature. He was adjunct lecturer in Environmental Conservation and Development at the University of Belize in 2007. He holds an MSc in Literature and Transatlanticism from the University of Edinburgh and BA in Biology and English Language and Literature from Goshen College in the United States.

Dina El Dessouky received her PhD in Literature at the University of California, Santa Cruz, where she is currently a Lecturer in the Writing Program. An essay based on her dissertation, titled *Identity and Island Place in Contemporary Kanaka Maoli and Ma'ohi Literatures*, appeared in the collection *Postcolonial Ecologies: Literatures of the Environment*, edited by Elizabeth DeLoughrey and George B. Handley. From 2008 to 2013, she served as co-editor of the international not-for-profit editorial collective *Kurungabaa*, a journal of literature, history, and ideas from the sea. Her most recent publication is a personal critical essay in the anthology *Min Fami: Arab Feminist Reflections on Identity, Resistance, and Space* (2014, edited by Ghadeer Malek and Ghaida Moussa). Dina is a 2013 Voices of our Nations (VONA) writing workshop Alum.

Eóin Flannery is a reader in Irish Literature in the Department of English and Modern Languages at Oxford Brookes University in the UK, where he also is Director of the Oxford Brookes Poetry Centre. He is the author of three books: *Colum McCann and the Aesthetics of Redemption* (2011); *Ireland and Postcolonial Studies: Theory, Discourse, Utopia* (2009); and *Versions of Ireland: Empire, Modernity and Resistance in Irish Culture* (2006). His next book, *Ireland and Ecocriticism: Literature, History, and Social Justice* is forthcoming in 2015. He recently edited a special issue of *The Journal of Ecocriticism* on "Ireland and Ecocriticism" (July 2013).

Adrian Taylor Kane is associate professor of Spanish and chair of the Department of World Languages at Boise State University. He is the editor of *The Natural World in Latin American Literatures: Ecocritical Essays on Twentieth Century Writing* (2010) and author of *Central American Avant-Garde Narrative: Literary Innovation and Cultural Change, 1926–1936* (2014). His articles and book reviews have appeared in such publications as

Bulletin of Spanish Studies, Interdisciplinary Studies in Literature and Environment, and *Review: Literature and Arts of the Americas*.

Priya Kumar is a specialist in postcolonial studies in the English Department at the University of Iowa, where she is an associate professor. She is the author of *Limiting Secularism: The Ethics of Coexistence in Indian Literature and Film* (2007). Her current work focuses on postcolonial narratives of mass displacement and exile, emphasizing the liminal figure of the refugee.

James McElroy, a native of Belfast, currently teaches at the University of California, Davis. He previously taught at the National University of Ireland, Maynooth, and Manhattan College, New York. His articles and reviews have appeared in *The New York Times*, *The Washington Post*, and *The Los Angeles Times*. The author of *Ireland: A Traveler's Literary Companion*, he has also published on ecocriticism and Irish poetry, Irish nature writing, and the Eco-Gothic.

Augustine Nchoujie recently completed his PhD in African literature at the University of Younde 1 in his home country of Cameroon. He has served as a visiting scholar at York University in Toronto, Canada, where he currently lives. He is a specialist in African environmental literature and ecocriticism.

Senayon Olaoluwa is a Wits University associate researcher and coordinates the Programme in Diaspora and Transnational Studies at the Institute of African Studies at the University of Ibadan in Nigeria. His papers and reviews have appeared in such journals as *English Studies*, *Journal of Postcolonial Writing*, and *Journal of African Studies*. His research interests include Diaspora, transnationalism, ecocriticism, postcolonialism, and cultural and film studies.

Zahra Parsapoor is assistant professor at the Institute for Humanities and Cultural Studies, Iran. She received her PhD in Persian Language and Literature from Tehran University. She has edited two collections of ecocriticism in Persian, and her ecocritical work has appeared in the issue of *Ravenshaw Journal of Literary and Cultural Studies* devoted to Green Studies (January 2014).

Swarnalatha Rangarajan has been associate professor of English at the Department of Humanities and Social Sciences at the Indian Institute of Technology Madras since 2010. She was a Fulbright Pre-doctoral fellow at Harvard University in 1999–2000. She was the founding editor of the *Indian Journal of Ecocriticism* (IJE) and has served as Guest Editor for two special issues on Indian ecosophy for *The Trumpeter*—the Canadian Journal of Deep Ecology. She has published articles on ecocriticism in several journals. In 2013, she held the Charles Wallace Fellowship at CRASSH (Cambridge University) to pursue a book project on ecocriticism. Her short fiction has appeared in anthologies from publishing houses like Penguin, Zubaan, and Westland, to name a few. She has also completed her first novel, which has a prominent ecosophical theme. She served as coeditor of *Ecoambiguity, Community, and Development: Toward a Politicized Ecocriticism* (2014).

Vidya Sarveswaran works as an assistant professor of English in the Department of Humanities and Social Sciences at the Indian Institute of Technology Jodhpur. She was a Fulbright Fellow for the year 2008–2009 at the University of Nevada, Reno. She completed her undergraduate and MA programs at the University of Madras and her doctoral degree at IIT Madras. She was the student editor of the *Indian Journal of Ecocriticism,* has several journal articles to her credit, and has worked on several documentary films as a script writer. She is interested in creative writing, film studies, and environmental humanities. She served as coeditor of *Ecoambiguity, Community, and Development: Toward a Politicized Ecocriticism* (2014).

Scott Slovic served as founding president of the Association for the Study of Literature and Environment (ASLE) from 1992 to 1995 and has edited the journal *ISLE: Interdisciplinary Studies in Literature and Environment* since 1995. The author of many books and articles in the field, his most recent publications include the coedited volumes *Currents of the Universal Being: Explorations in the Literature of Energy* and *Numbers and Nerves: Information, Emotion, and Meaning in a World of Data,* among other projects. He is professor of Literature and Environment and chair of the English Department at the University of Idaho. He served as coeditor of *Ecoambiguity, Community, and Development: Toward a Politicized Ecocriticism* (2014).

Anthony Vital, a native of South Africa, is professor of English at Transylvania University in Kentucky. He began presenting conference papers on postcolonial ecocritical approaches to South African literature in the late 1990s, and his research on ecocriticism and African literature has appeared in numerous books and journals, including *Environment at the Margins: Literary and Environmental Studies in Africa, Safundi: The Journal of South African and American Studies, Research in African Literatures,* and *Journal of Commonwealth and Postcolonial Studies,* among others.

Munazza Yaqoob is assistant professor and chairperson in the Department of English, Female Campus, at the International Islamic University in Islamabad, Pakistan. She has published previous articles on classical and Romantic poetics, feminist pedagogy, and ecocritical readings of Pakistani literature, and her research interests also include comparative literature and literary theory.

Zhou Xiaojing is Director of the Ethnic Studies Program and professor of English at the University of the Pacific in Stockton, California. A native of the People's Republic of China, she has published a number of articles on American and international environmental literature and she also publishes Chinese translations of American ethnic poetry in Chinese journals.